KEY THEMES IN GEOGRAPHY

Urban and Rural Geography

Jack Gillett

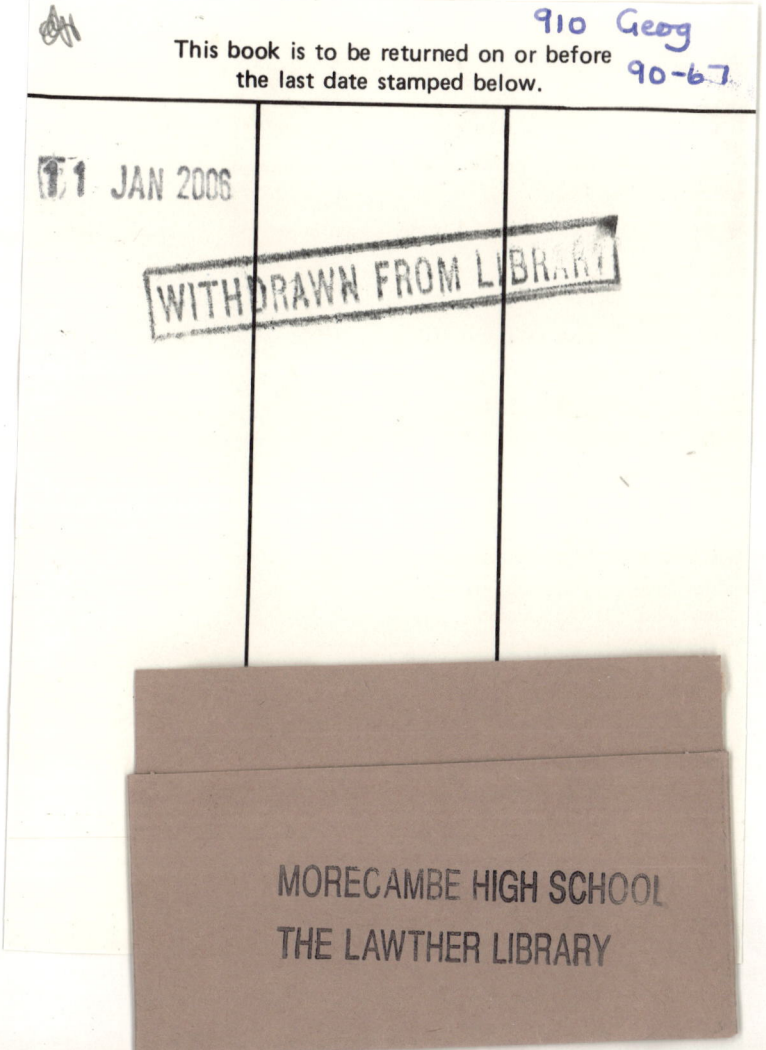

Longman Group UK Limited,
Longman House, Burnt Mill, Harlow,
Essex CM20 2JE, England
and Associated Companies throughout the world.

First published 1988

Set in 10/12 point Palatino (Linotron)

Produced by Longman Group (FE) Ltd
Printed in Hong Kong

ISBN 0 582 23176 0

British Library Cataloguing in Publication Data

Gillett, Jack
   Urban and rural geography.
   1. Geography —— Text-books —— 1945–
   I. Title
   910   G128

For Trev. and Eileen Birdsall

We are grateful to the following for permission to redraw artwork:

Allen and Unwin, *Elements of Human Geography*, C. Whynne-Hammond, fig 4.11; Automobile Association, fig 5.1; *Daily Telegraph*, page 107 and page 147; *The Geographical Magazine*, October 1972, fig 5.16; Oxford University Press, *The Moscow City Region*, F.E.I. Hamilton, fig 3.24; Pluto Press, *The State of the Nation*, figs 2.45 and 2.46; Warrington and Runcorn Development Corporation, figs 5.12 and 8.23; World Bank, *The Development Data Book*, fig 1.4.

We are also grateful to the following for permission to reproduce photographs and other copyright material:

Automobile Association, fig 7.3; Aerofilms, fig 2.27; Airviews, Manchester, figs 2.9, 5.9, 8.14; Heather Angel, fig 10.20; Arcaid/Lucinda Lambton, fig 2.13a; BBC Hulton Picture Library, figs 1.22, 3.22; Camera Press, figs 2.32, 6.9; J. Allan Cash, figs 1.12, 2.12a, 3.12, 3.15, 3.16, 4.2, 4.6, 4.12, 10.5, 10.6, 11.6; Celtic Picture Agency/Barry Webb, fig 2.11a; *Daily Telegraph*, fig 2.28; Douglas Dickins, fig 3.8, 5.14; Earthscape, figs 1.7, 1.11, 1.14, 1.21, 3.11, 3.20, 4.15, 5.17, 7.2, 7.8, 10.4, 10.7, 10.23, 11.2; Mary Evans Picture Library, fig 2.3; Forestry Commission, fig 10.18; Jack Gillett, figs 1.25, 1.26; Sally & Richard Greenhill, fig 4.4; Harland & Wolff, fig 4.10; John Hillelson Agency/R. Singh, fig 5.15; Irish Linen Guild, fig 4.7; Peter Lake, fig 8.3; Leeds City Libraries, fig 2.4; Link Picture Library, fig 4.18 (Greg English), 4.19; Donald McLaren, fig 6.6; Manchester City Council, fig 2.10; Ministry of Agriculture, Fisheries and Food, Crown Copyright, permission to use this advertisement implies no endorsement of the views held in this book, fig 10.16; Mitchell Library, Glasgow, fig 2.5; David Muscroft, fig 6.15; The National Trust, fig 11.1; Network/Katalin Arkell, fig 3.6; Oxfam, figs 6.1, 6.3, 6.4 (Melvin Almond); Page & Wells, fig 12.5; Panos Pictures, figs 3.3 (Glen Edwards), 3.5 (Mark Edwards), 6.7 (Sean Sprague); Photosource/Keystone, fig 3.1; Picturepoint, figs 1.13, 3.2, 3.4, 3.9, 3.18, 3.23, 4.3; Popperfoto, figs 4.8, 8.5 (Associated Newspapers), 10.21; Reflex Picture Agency/Dennis Doran, fig 6.5; Sheffield Newspapers, fig 10.13; Topham, figs 2.30, 10.22 (PA); Tyne & Wear Transport, fig 7.7; UK Tobacco Company, fig 5.8; Warrington & Runcorn Development Corporation, fig 8.19

Photographs for figs 12.4 and 12.6 were taken by J. A. Belsom. All other photographs taken by Brendan Hearne.
The Ordnance Survey map extracts (figs 2.11b, 2.12b, 2.13b, 12.1, 12.2) are printed with the permission of the Controller of Her Majesty's Stationery Office, Crown Copyright reserved.

The cover photograph of Hundertwasser's house in Vienna is by Erich Lessing/Magnum/John Hillelson Agency.

Contents

Preface

The Longman *Key Themes in Geography* Series meets the requirements of many of the GCSE syllabi introduced in 1986. Teachers involved in Schools Council (Avery Hill/GYSL) based courses should find this series particularly helpful as its topics have been grouped together under economic, environmental and urban/rural headings.

The books in the series have been designed to enable students to achieve the high academic aims specified in the GCSE National Criteria. Emphasis has been placed on practical work and enquiry-based techniques, as these inevitably bring students into the closest possible contact with the subject material. Ample opportunities for individual research and group role-play have been built into the teaching programmes, and each book contains a selection of both aim and hypothesis-based fieldwork topics to meet course study requirements. Additional guidance and recommended activities for this vital component are provided by *Fieldwork Studies in Geography* (Longman; ISBN 0 582 22442 X).

The teaching units are designed to convey basic concepts, knowledge and skills; all units include a set of questions which are graded in difficulty. Some of these questions invite class discussion and are deliberately 'open-ended'; this allows the teacher to dictate how much information the students are expected to put into their answers. The three books in the series each end with a test/revision section – also carefully graded.

Topics are examined at local, regional/national, and international/global scales, as required by the National Criteria. An appendix to each book carries a world map on which studies at the last scale are located. A further appendix in each book provides a glossary of key terms; these are highlighted in the text by bold type.

The authors welcome constructive comments on the books in this series, and ask that these be directed through the publishers.

Jack Gillett
Series Editor

Broughton, Preston
September, 1988

Companion Volumes
Economic Geography (ISBN 0 582 23177 9)
Environmental Geography (ISBN 0 582 23178 7)

1

Introducing Urbanisation

1.1 Introduction: quality of life

Quality of life* is one of the most important ideas in modern geography, and it features in many of the units in this book. Put very simply, quality of life means the general state of an area and the well-being of the people living in it. As Fig. 1.1 shows, it includes almost every aspect of our lives. You can probably think of other aspects which are not included in this cartoon.

1 Copy out and complete this sentence: 'Quality of life describes the general ... of an area, and the well-being of its ...'.

2 Assess the quality of life of your own local area by:
a Discussing openly in class each of the quality of life 'themes' shown in Fig. 1.1.
b Writing sentences about at least six of these aspects to make *your own* assessment of the quality of life in the local area.

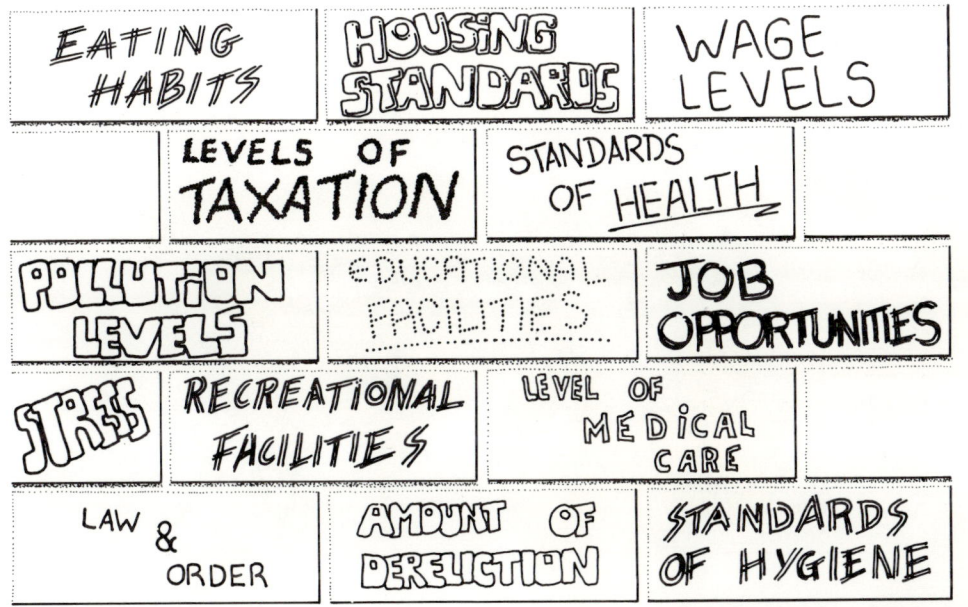

EATING HABITS

HOUSING STANDARDS

WAGE LEVELS

LEVELS OF TAXATION

STANDARDS OF HEALTH

POLLUTION LEVELS

EDUCATIONAL FACILITIES

JOB OPPORTUNITIES

STRESS

RECREATIONAL FACILITIES

LEVEL OF MEDICAL CARE

LAW & ORDER

AMOUNT OF DERELICTION

STANDARDS OF HYGIENE

* Note: the meanings of all the key words printed in bold type are given in the glossary (pages 153–156).

◄Figure 1.1 'Quality of life' is made up of many things

▼Figure 1.2 'Delightful' house in the countryside

Everyone has their own idea of 'the perfect life' – the life-style which they feel would make them completely happy. For some it is a large detached house set in its own beautiful grounds (Fig. 1.2), together with a host of the latest gadgets to make life easier and more enjoyable. For *many more*, it is enough food to eat, a healthy body, a job (almost any kind will do), a sturdy weather-proof shelter to live in and perhaps a bicycle for getting to work. These are, of course, two extremes but they do help us to appreciate how widely quality of life varies throughout the world at the present time.

1.2 Developed and developing countries

Figures 1.3 and 1.4 show how literacy rates and life expectancy vary throughout the world. These are just two aspects of quality of life, and additional maps could also have been drawn for many others (e.g. the numbers of cars, doctors or telephones per 1000 people). In fact, many of these additional maps would look very similar to those on the opposite page. Now try to answer Questions 1 to 3 – which are based on these maps – before reading the rest of this unit.

1 **a** Which of the continents named in Fig. 1.3 have the highest literacy rates over most of their area? (i.e. over 75%)
b Which of the continents named in Fig. 1.4 have the longest life expectancy over most of their area? (i.e. 70 years or more)
c What do your answers to a and b have in common?

2 Write the name of each of these countries (Algeria, Bolivia, Canada, Chad, Indonesia and Zaire) twice to complete your copy of the table below. Figures 1.3 and 1.4 have all the information you need to do this.

Countries having these literacy rates:	Countries having these life expectancies:
0–24%:	< 50 years:
25–49%:	50–59 years:
50–74%:	60–69 years:
75–100%:	≥ 70 years:

3 Which of these statements are correct, according to Figs 1.3 and 1.4?

'*Countries which have high literacy rates usually have long life expectancies also.*'

'*Countries which have lower literacy rates usually have shorter life expectancies also.*'

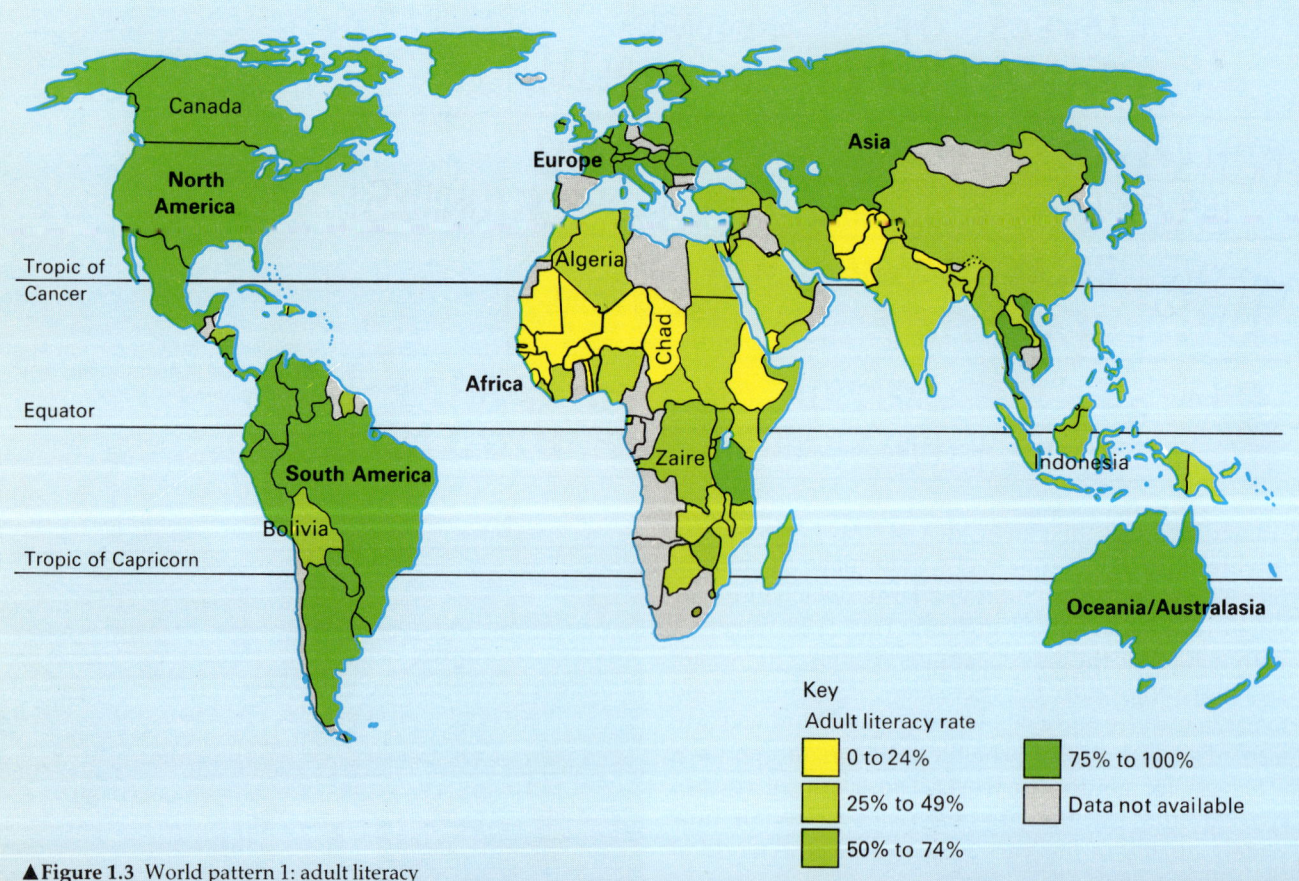

Key
Adult literacy rate
- 0 to 24%
- 25% to 49%
- 50% to 74%
- 75% to 100%
- Data not available

▲ **Figure 1.3** World pattern 1: adult literacy

'All the countries with high literacy rates and long life expectancies lie outside the tropics; this means that they are either north of the Tropic of Cancer or south of the Tropic of Capricorn.'

'Many of the countries with lower literacy rates and shorter life expectancies lie between the Tropics of Cancer and Capricorn; some of them are actually on the Equator.'

Gross National Product

It is clear that some countries enjoy a much higher quality of life (standard of living) than others. These countries are called **developed countries**, and many – but not all – of their people have an adequate life-style. Only one-quarter of the world's total population live in these countries.

The other 75% of the world's people inhabit (live in) **developing countries**. They are sometimes referred to in newspapers and on the television as 'less developed' or 'underdeveloped' countries. These poorer countries belong to the **Third World** (the 'First World' includes developed, **capitalist** countries such as Australia, France and the United Kingdom; **communist**, developed countries like the Soviet Union belong to the 'Second World').

Deciding whether a country is rich enough to be called developed is easier than you might think. It is usually done by taking the *total* income for the whole country over a full year (its **Gross National Product**, or GNP for short), then dividing this figure by the country's population – the number of people who have to share this income. The result of this simple calculation is called the **per capita GNP**; the higher this figure is, the richer a country must be. There are two important points to remember about GNP and per capita GNP:

□ The GNP figures for some Third World countries are unreliable. This is because their people cannot or do not feel the need to keep records of their business activities.

□ Per capita GNP is only an *average* figure, for a country as a whole. It cannot show what proportion of the national wealth each person enjoys, and so may hide great differences in quality of life within a country.

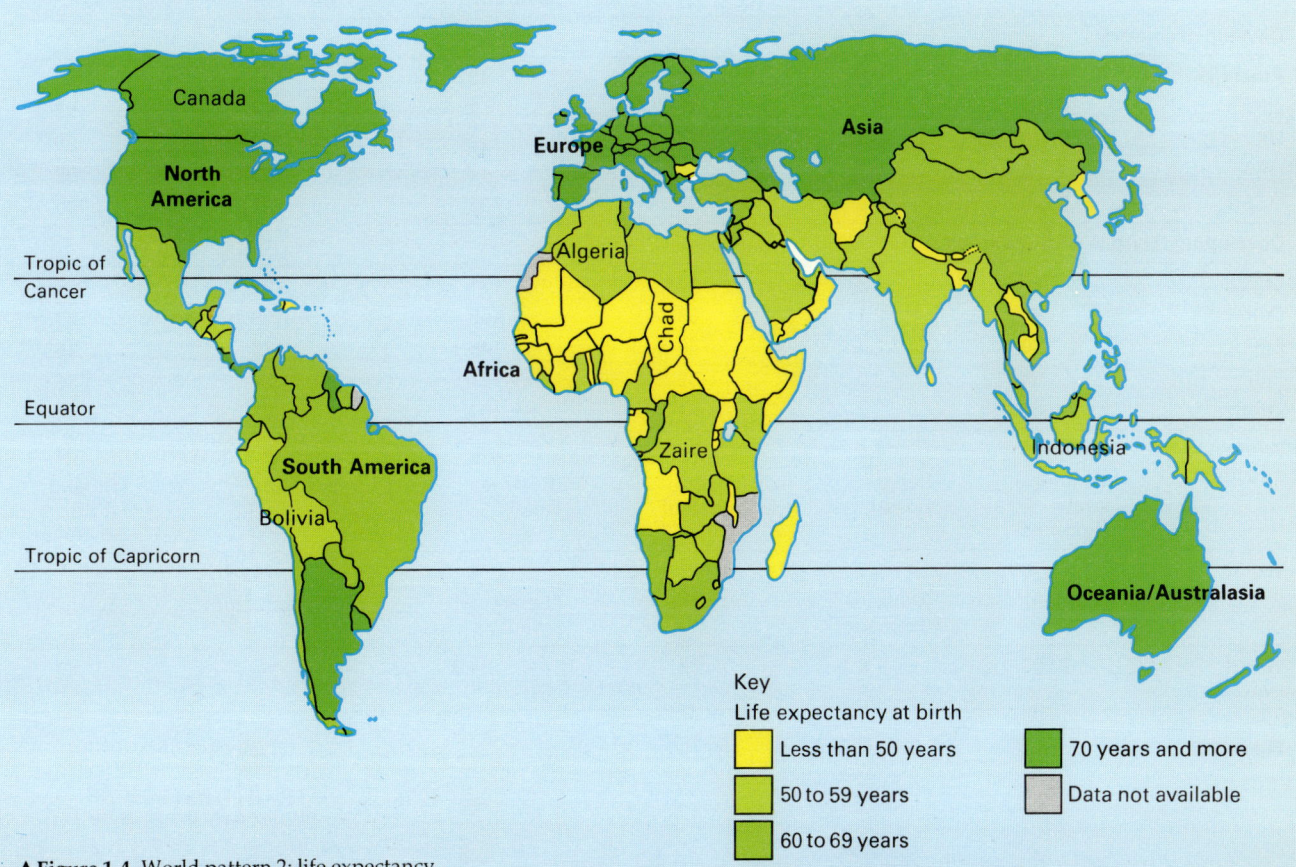

▲**Figure 1.4** World pattern 2: life expectancy

Key
Life expectancy at birth

- Less than 50 years
- 50 to 59 years
- 60 to 69 years
- 70 years and more
- Data not available

Figure 1.5 lists six basic factors which, when taken together, give us some idea of the quality of life in each of these countries. This table also includes the maximum, minimum and average figures of these factors, for the world as a whole.

It is wrong to think that Third World countries can *never* become developed. Brazil, Hong Kong and South Korea are all relatively poor at the moment, but their economies are growing four times as rapidly as our own. We expect their per capita GNPs to rise quite quickly in the near future, but this will happen only if their GNPs grow faster than their populations. (Remind yourself of the definition of per capita GNP if you can't understand why this should be.) As the third column of Figure 1.5 shows, the developing countries tend to have an above-average rate of **annual population growth**. For example it will take only 22 years for the Ivory Coast's population to double, compared with 533 years for the United Kingdom, and 40 for the whole world.

4 a Copy out and complete this table by writing the per capita GNP figures for the five countries in the last column; these may be found by dividing their GNPs by their populations.

Country	GNP in US$	Population	Per capita GNP in US$
Belgium	119 200 million	10 million	
Canada	273 600 million	24 million	
Chile	28 160 million	11 million	
Ghana	4 800 million	12 million	
Sri Lanka	4 500 million	15 million	

	Per capita GNP (US$)	Annual population growth (%)*	Life expectancy (years)	Infant mortality (deaths under 1 year per 1 000 children)	Daily intake of calories (as % of needs)	Literacy rate (%)
Developed countries						
Australia	11 080	0.8	74	10	117	100
Japan	10 080	0.7	77	7	124	99
UK	9 110	0.1	74	12	132	99
USA	12 820	0.7	75	12	139	99
USSR	5 940	1.0	72	32	132	100
Developing (Third World) countries						
China	300	1.3	67	71	107	69
Colombia	1 380	2.1	63	55	108	81
Ethiopia	140	2.4	46	145	76	15
India	260	2.0	52	121	87	36
Zambia	600	3.2	51	104	93	44
World maximum	24 660 (United Arab Emirates)	4.6 (Ivory Coast)	77 (Japan, Sweden)	208 (Burkina Faso)	160 (Belgium)	100 (Finland, USSR, Australia)
World minimum	80 (Bhutan)	−0.1 (West Germany)	42 (Angola)	7 (Japan, Sweden)	70 (Mozambique)	5 (Burkina Faso)
World average	2 500	1.8	62	84.0	n/a	68

Figure 1.5

* The figures in this column ignore the movements of people to and from other countries; they are based purely on the birth and death rates of permanent residents.

b Which of these are likely to be developed countries (see Fig. 1.5)?

c Use the index in your atlas to find out which continent each of these countries is in.

5 With the help of Fig. 1.5, and any other factors which you think are important, describe the main differences between developed and developing countries. It might be helpful to discuss your ideas openly in class first.

Population change

Figure 1.6 demonstrates that the developing countries have a higher rate of annual population growth because there is a wider gap between their **birth rates** and **death rates**. The death rate, at first very high, is greatly reduced by better medical facilities. Originally, the birth rate was also very high – to allow for the high death rate! Parents are now beginning to realise that most of their children will survive, but it will take some time for everyone to be convinced of this. Many people in the Third World still prefer to have large families as the children can help on the land and look after their parents when they are old. The fathers regard many children as a status symbol.

6 Study the table below, which shows the birth and death rates (per 1000 people) for the same five countries.

Country	Birth rate	Death rate
Belgium	13	11
Canada	15	7
Chile	22	7
Ghana	48	16
Sri Lanka	28	6

a In which of these countries is the population increasing:
  most rapidly?
  most slowly?
b Explain how the annual population growth figure is important to a country's economic growth and quality of life.

7 a Copy out and complete this summary of the information shown in Figure 1.6. Use the words *annual population growth, birth* and *death* to fill the blank spaces.

  The annual population growth rate usually changes as a country becomes wealthier and more developed. This is because its birth and death rates are also changing.
  The poorest countries have a high ... rate as well as a high ... rate. This means that their ... rate is quite small.
  Many developing countries are at the next stage. Their ... rate remains high, but the ... rate falls sharply. Because of this, their ... rate is increasing rapidly.
  As these poorer countries move towards the developed stage, their ... rate remains low and the ... rate falls to about the same level; the ... also decreases as this happens.
  The countries which have been developed for many years have ..., ..., and ... rates which are all quite low, and so their populations are fairly stable; they may even decrease from time to time.

b After class discussion, state why the trends shown in Fig. 1.6 mean that the world could avoid becoming dangerously overcrowded.

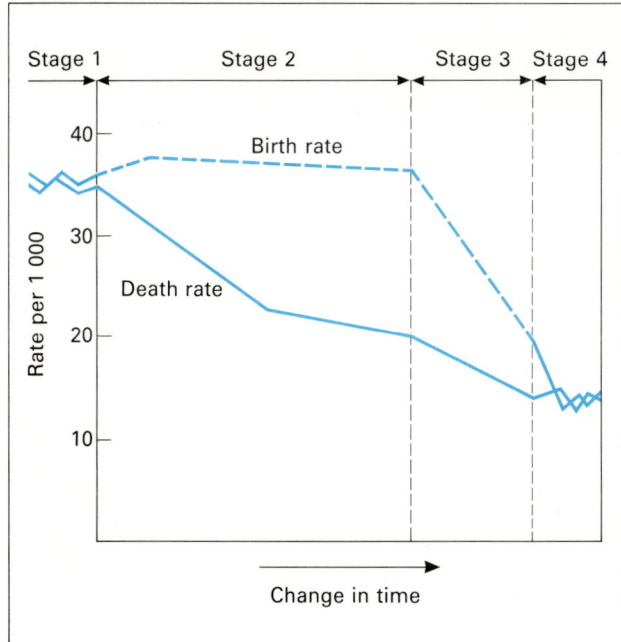

▲**Figure 1.6** The four stages of population growth which countries pass through as they become richer. The growth rate is shown by the gap between the birth and death rates. Most developing countries are at Stage 2; Britain is at Stage 4

◀ **Figure 1.7** Oil has transformed life in Kuwait

▼ **Figure 1.8**

Some Third World countries have been very fortunate because they have discovered large reserves of oil. The wealth earned by exporting the oil has boosted their per capita GNPs so much that they are now amongst the highest anywhere in the world. Oil-rich Kuwait for example has modern roads, a reliable water supply, superb medical facilities and a totally free telephone service! Most of the developing countries have to rely on trading with the richer countries and any aid they receive from them. Unfortunately, these links can make the developing countries very dependent on the richer ones and may even restrict their own industrial growth.

Quality of life varies a great deal not only between countries, but *within* them as well. The variation between the life-styles of the rich and the poor is especially marked in Third World countries. Even in developed countries such as Britain, there are quite striking contrasts between whole regions and the different **groups of people** living in them. Governments often give special help to their poorer regions, but rarely enough to remove the stubborn differences (e.g. in unemployment) which have built up over many years.

8 Pair up these important terms with their correct meanings.

A *developed country*
A *developing country*
Annual population growth
Gross National Product
Per capita GNP

... is a country's annual income divided by its population.

... is a fairly wealthy country which can afford to meet the needs of many of its people.

... is a generally poor country which is part of the Third World.

... is the difference between a country's birth rate and its death rate.

... is the total yearly income of a country.

9 **a** Explain why Kuwait's GNP has risen sharply in recent years.
b Copy the sketch of the scene in Fig. 1.7 then add labels to the things in the scene that are evidence of modern construction and wealth.
c Note down any *other* ways in which Kuwait's new wealth has enabled that country to improve its quality of life.

1.3 Migration to the cities

One striking feature of modern times has been the **migration** (movement) of large numbers of country-dwellers into towns and cities. The process by which an increasing *proportion* of a country's population lives in large settlements is called **urbanisation**. Figure 1.9 shows that on average, two out of every five people in the world now live in towns and cities; also that the *rate* of urbanisation throughout the world has increased most rapidly since about 1950.

Figure 1.9 is of course looki[...]
whole. It can therefore only gi[...]
what has taken place over th[...]
1.10 is much more helpful because i[...]
isation trends in two contrasting countries. [...]
rise for the United Kingdom, a typical developed country, was due to the creation of jobs in the booming industrial towns of the nineteenth and early twentieth centuries. The pattern of urbanisation in developing countries such as Ghana is quite different. Industrialisation came much later to these countries, but the main reason for their recent very high rate of urbanisation is the serious lack of work – and often food too – in their **rural** (country) areas.

There are many reasons why people throughout the world have opted for city life. These reasons can be grouped into what are called **pull factors** and **push factors**. The pull factors are the advantages (good points) of a place which make it attractive to live in. The push factors are the disadvantages; they encourage people to move *away* from their home area. The most common examples of each type of factor are show on pages 12 and 13. Some of them are described in sentences, while others can be obtained from the illustrations (Figs 1.11–1.14).

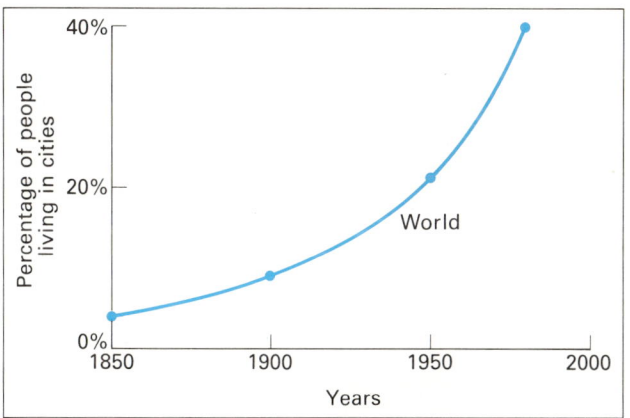

▲**Figure 1.9** The trend towards urbanisation for the world as a whole

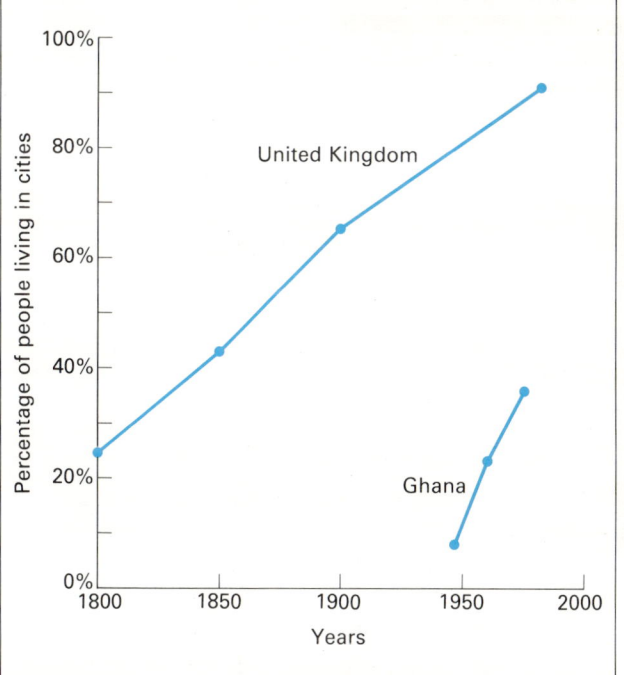

▲**Figure 1.10** Urbanisation trends in two contrasting countries

▲**Figure 1.11** Coffee bean washer at a factory in Kenya

◄ **Figure 1.12** A government housing block in Nigeria

◄ **Figure 1.13** Malaysia Hotel, Singapore

▼ **Figure 1.14** Transporting water in Kenya

'The soil may become exhausted (worn-out) because the farmers cannot afford to fertilise it properly. Infertile land produces less food each year.'

'Much of the farmland in the developing countries has been "fragmented" (divided up between a farmer's sons when he dies). Over many generations of farming families, these parcels of land become too small to feed their owners.'

'The population of a farming area may grow so rapidly that the land there can no longer produce enough food for everybody.'

'Few rural areas in developing countries have electricity supplies, safe drinking water and adequate sewage disposal facilities.'

'There are greater promotion prospects in the towns and cities. City-dwellers do at least have a chance to better themselves and achieve some of their ambitions.'

'Traditional farming methods usually mean a lot of hard work. Most farmers in the developing countries are "subsistence farmers" (people who try to grow enough food for their families, but cannot guarantee to produce a profit-making surplus). This kind of farming doesn't encourage young people to stay in agriculture.'

'The best schools are usually found in the largest settlements, and they often attract the most highly qualified teachers. The universities and colleges are also located there.'

'Large "multi-national" companies now operate worldwide and are buying farms to produce the food they sell. This increases their profits, but puts the original landowners out of work. Many of them go to the cities in search of new work.'

There is nothing particularly new about most of these pull and push factors. What *has* changed is the ease with which people in country areas can find out what the cities can offer them. Television, radio and newspapers now reach most rural areas, even in the poor countries of the world. They usually make city life look exciting and rewarding, and often ignore its less attractive aspects.

Urbanisation has also been encouraged by recent improvements in public transport. These reduce travelling times to the cities and give migrants (the people who move) the security of knowing that they may be able to return to their villages if city life proves disastrous for them.

Most of the information in this unit has been written with the developing countries in mind. It is a sad fact that hundreds of millions of people in these countries still live in such poverty that they see a move to a city as their only hope for survival. Many of them must leave their villages knowing full well that their new life will be extremely hard; to stay, however, is unthinkable.

Some of the factors which have been described and illustrated also apply to developed countries. Their greater wealth has improved the quality of life in *both* rural and urban areas. Migration from one to the other in these countries is largely a matter of personal choice rather than absolute necessity, and many of the wealthier inhabitants of their cities now prefer to live in or near open countryside. Chapters 2 and 8 discuss some of the serious effects of this 'counter-urbanisation' on urban and rural areas in the developed countries.

1 Copy out this paragraph then write *migrants, pull factors, push factors, rural* or *urbanisation* in the blank spaces.

... means that the populations of towns and cities are growing faster than the populations of ... (country) areas. Millions of ... throughout the world are now moving into the cities, spurred on by their attractions (...), and the problems (...) of country life.

2 **a** Discuss what Figs 1.11–1.14 are trying to show.
b Draw a table with these three headings. The first column must be made much wider than the other two.

Description of factor	Pull factor	Push factor

c Using Figs 1.11–1.14 and the tinted text complete your table by writing a *brief* summary of each factor in the first column; use ticks to complete the other two columns.
d Total the number of ticks in each column. What do these totals appear to tell us about rural → urban migration? Do *you* feel these totals are a reliable source of information? Give reasons for your last answer.

1.4 Millionaire cities

Urbanisation has greatly increased the number of very large cities in the world. Figure 1.15 shows the total number of **millionaire cities** at ten year intervals since 1920; these are cities with at least 1 000 000 inhabitants. For information, there was *only one* millionaire city in 1800– and that was London.

The millionaire cities in the developing countries have proved so attractive to country-dwellers that many of them are growing at an astonishing rate. Mexico City for example had 370 000 inhabitants in 1900. Its current population is about 15 million, and this is expected to rise to a staggering 36 million by the year 2000. Places like Mexico City would have grown anyway, because the total populations of Third World countries are increasing quite rapidly at the present time. It is the *migration* of people from rural areas which has caused the city populations to grow much faster than those in rural areas.

Year	Number of millionaire cities in the world
1920	24
1930	39
1940	41
1950	80
1960	113
1970	149
1980	181

Figure 1.15 The increasing number of millionaire cities

Conurbations

Some large cities have **sprawled** outwards to produce huge built-up areas called **conurbations**. These often include smaller towns and villages which were once in open countryside but are now part of an almost continuous mass of **urban** development. Britain has seven major conurbations – all of which have well over one million people.

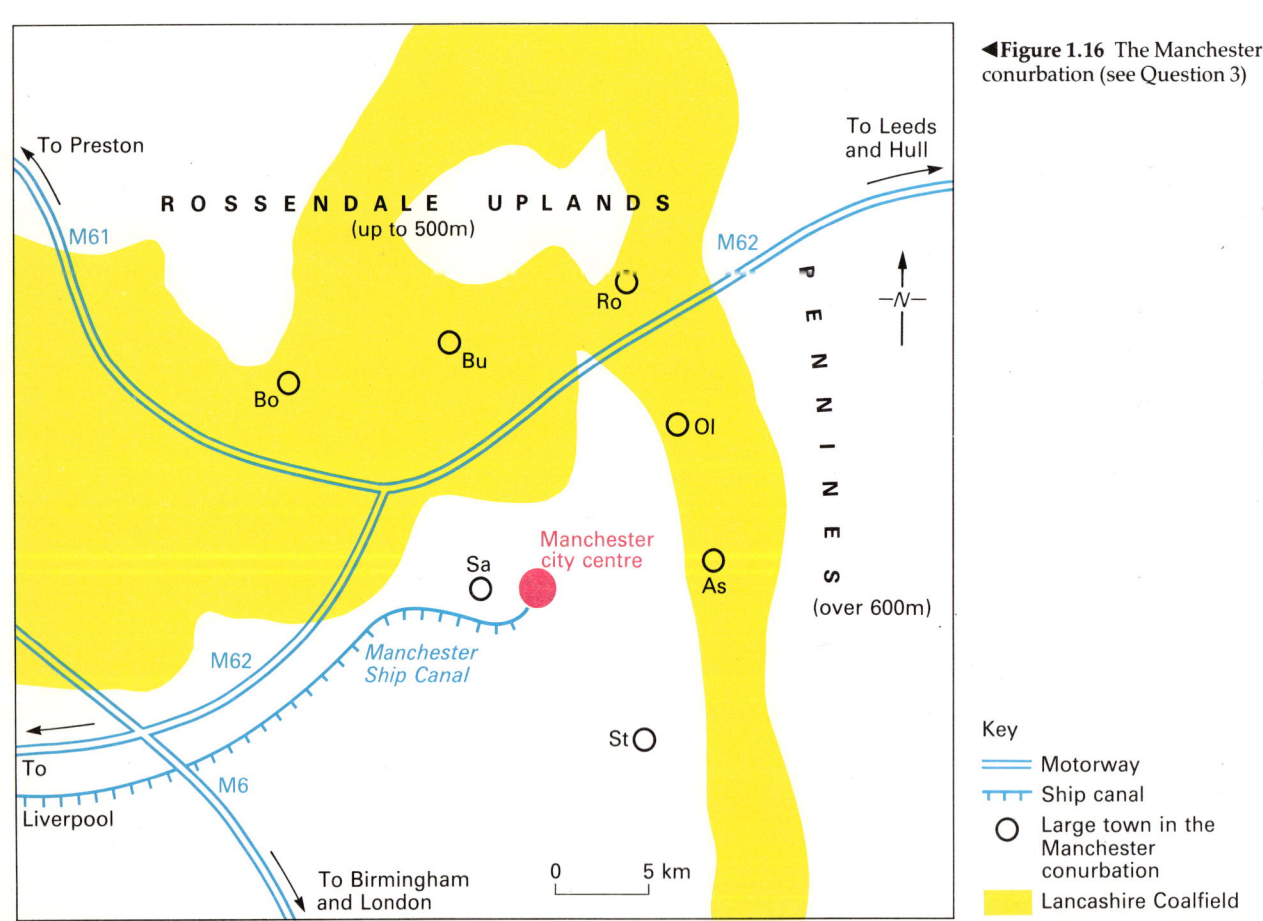

◀**Figure 1.16** The Manchester conurbation (see Question 3)

APPALLACHIAN MOUNTAINS

(rising above 1 000m)

Hudson River

Boston

Providence

New York

Long Island

Philadelphia

Baltimore

Washington

ATLANTIC OCEAN

Chesapeake Bay

Richmond

Norfolk

0   100km

Key
■ Capital city
● Other large cities
▢ Megalopolis area

—N—

Hokkaido

Sapporo

Sea of Japan

Honshu

Mount Fuji

This is the largest area of lowland on Honshu

Tokyo

Yokohama

Kyoto Nagoya

Kobe

Hiroshima

Osaka

Fukuoka

Shikoku

Inland Sea

Kyushu

PACIFIC OCEAN

0   200km

▲ **Figure 1.17** The North American megalopolis
▶ **Figure 1.18** The Japanese megalopolis

Figure 1.16 shows the layout of the Manchester conurbation. This area became important as a centre of the cotton textile industry during the eighteenth and nineteenth centuries. Its present population is 2 600 000, but only 449 000 of these people live in the 'city' of Manchester. The remaining 83% inhabit the other settlements which are now part of the conurbation, and some of them have as many as 150 000 people. Only the first two letters of their names are shown on this map because Question 3 b invites you to write down their full names – with the help of an atlas map of North-West England.

Megalopolis conurbations

Some conurbations abroad are so large that they have been given a special name: **megalopolis**. The two outstanding examples are on the east coast of the United States (Fig. 1.17) and of Japan (Fig. 1.18). The American megalopolis stretches for 800 km along the North Atlantic coast. Migrants have been attracted by its many industries and the capital city (Washington, D.C.). Its population has now reached 638 000, but New York and Philadelphia are much larger, with 7 million and 1 688 000 people respectively.

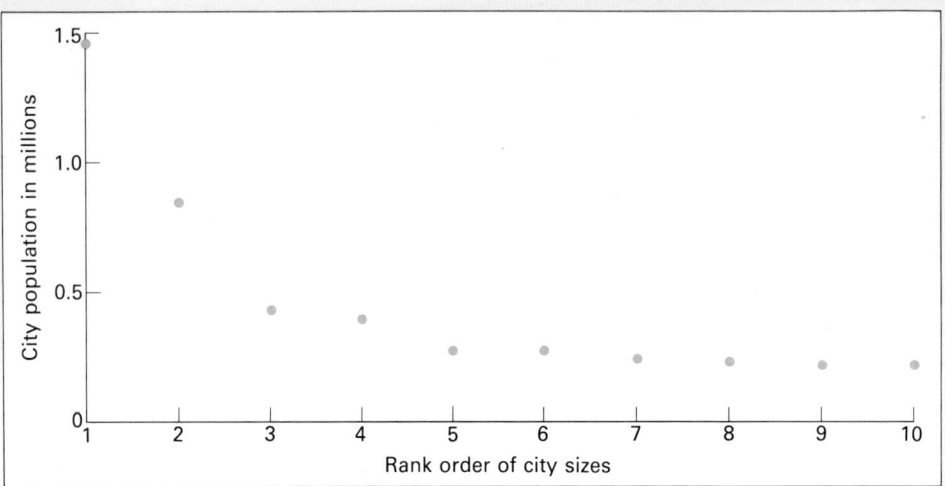

◄Figure 1.19 Scatter graph showing the populations of Nigeria's ten largest cities

Primate cities

Most countries include one city which is considerably bigger than its closest rivals (Fig. 1.19). These are called **primate cities**, and many of them are major ports or the capitals of their countries.

1 Explain the meanings of these important terms:

conurbation   primate city
megalopolis   urban
millionaire city   urban sprawl

2 **a** With the help of a calculator, work out the average percentage increase in population for each list. These figures show the percentage increase during 1965–80 for the world's five largest conurbations and cities in each group.

Cities/conurbations in developed countries		Cities/conurbations in developing countries	
New York	15%	Mexico City	113%
Los Angeles	86%	Shanghai	43%
Paris	16%	Buenos Aires	39%
Tokyo	69%	Beijing	71%
Moscow	13%	Calcutta	75%

b Write a sentence to compare your two average figures.
c With the help of an atlas, locate and name these ten cities on an outline map at the world. Use a different coloured dot for each group.

3 **a** Write down the main difference between a conurbation and a megalopolis.
b Copy Fig. 1.16 adding the full names of the towns shown on this map by just their first two letters.

c After class discussion, explain why over 2.5 million people now live in the Manchester conurbation. (Hint: consider location, relief, fuel, industry and transport.)
d Find the similarities (and differences) between the two megalopolis areas shown in Figs 1.17 and 1.18. You could split into small 'discussion groups' to prepare your answers to parts d and e of this question.
e Suggest the kinds of problems which a megalopolis might face due to its size. (Hint: start by considering traffic, pollution and quality of life.)

4 **a** Draw two scatter graphs (like the one in Fig. 1.19) to plot the information given in this table.

| Rank order of city size | Populations of the ten largest cities in: | |
	Algeria (Developing country)	Sweden (Developed country)
1 (largest)	1 500 000	1 380 000
2	485 000	694 000
3	350 000	454 000
4	313 000	143 000
5	224 000	120 000
6	159 000	118 000
7	157 000	117 000
8	151 000	111 000
9	128 000	108 000
10	115 000	103 000

b State whether your two dot patterns look alike.
c Try to explain *why* most countries are dominated by one city.

1.5 Urbanisation and rural areas

You now know that urbanisation has helped the world's cities to grow much larger. Our study of urbanisation would not be complete, however, without studying some of its effects on the *countryside* – the areas from which the migrants have come. In fact rural to urban migration has helped to reduce population pressures in overcrowded rural areas. Unfortunately, it has also produced some quite serious problems for these areas. This unit looks in detail at one such area – the Highlands and Islands region of Northern Scotland (Fig. 1.20).

This is one of Britain's most inhospitable (difficult) areas. Parts of it are mountainous, while others are very remote (a long way from the main towns). The whole region receives a high annual rainfall and its soils are much thinner and less fertile than those further south. Severe gales sweeping in from the Atlantic Ocean and heavy snowfalls in winter (Fig. 1.21) often make it very *inaccessible* (hard to get to). All these push factors have encouraged widespread emigration from this region.

Migration from Northern Scotland has however been only partly due to the 'natural' problems just described. There were *historical* reasons too and these meant that the early migration was far from peace-ful. In 1745, Bonnie Prince Charlie inspired an uprising of the Highland clans (groups of people having the same surnames). His aim was to invade England and remove King George II from the throne. The English responded by sending an army into Scotland, where it defeated the clans at the Battle of Culloden. This so weakened the clans that their chiefs could no longer force the Highlanders to stay in their villages. Some people left, reducing the chiefs' incomes and forcing them to consider other ways of earning money. They quickly realised that sheep could be highly profitable, and many chiefs were totally ruthless in evicting their own clansmen to make room for the new flocks (Fig. 1.22). The Highland Clearances, as this process was called, lasted from 1785 to 1850. Thousands of Highland families migrated to the booming industrial towns around the River Clyde where they found work in coal mines, textile mills and shipbuilding yards. Many also migrated abroad and started a completely new life in Australia, New Zealand and North America.

▼Figure 1.21 Highland winters can be harsh

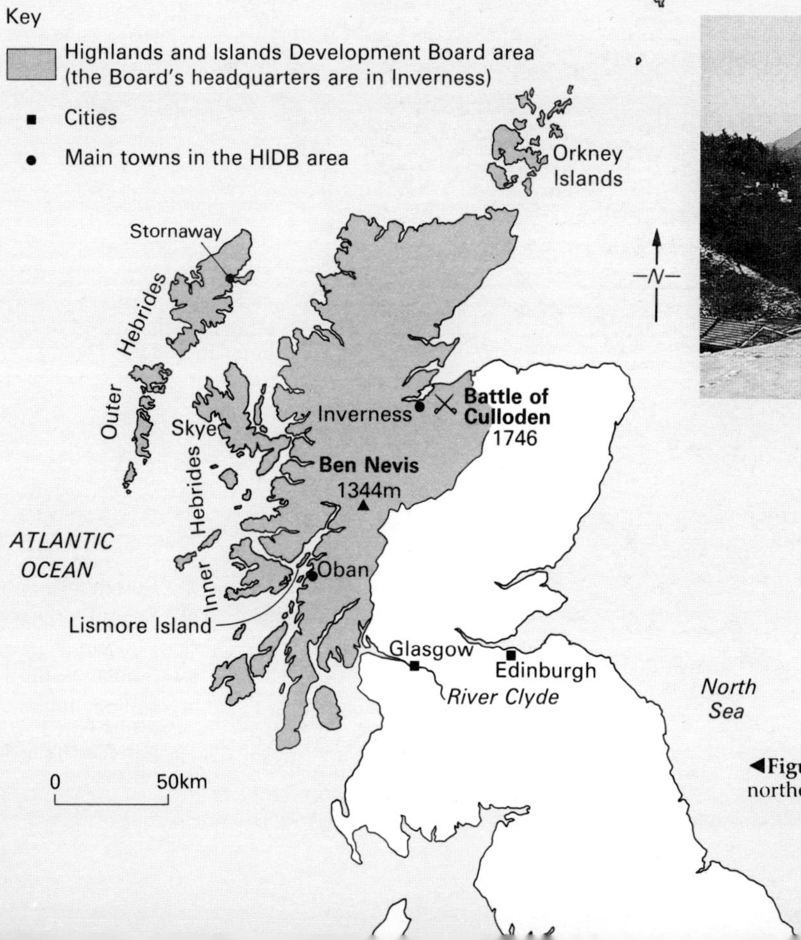

◀Figure 1.20 The Highlands and Islands region of northern Scotland.

The Highlands and Islands never really recovered from this mass-exodus of its people. Most of the emigrants were energetic young families with children, and some are still leaving certain parts of the region. Between 1961 and 1981, no fewer than 27 Scottish islands became deserted due to migration. This was one-fifth of the number which were inhabited in 1961.

One Scottish island which has lost many people but is still inhabited is Lismore (Figs 1.20 and 1.23), in the Inner Hebrides. Its highest population, recorded in the 1831 **census**, was 1790 (Fig. 1.24). These people must have led had a difficult life, for the island is only 28 km^2 in area and could not have supported such a large farming community. Today, its population is only 156, less than 9% of the peak total. Lismore farmers now complain that there aren't enough people to work the land properly!

Evidence of this migration can be seen throughout the island. Lismore is littered with ruined, abandoned dwellings (Fig. 1.25) and 24 of its houses are 'holiday homes'. These are owned by visitors, who occupy them for only a few weeks each year. Port Ramsay was once a busy farming and fishing community. Now most of the white-washed cottages along its one and only street are holiday homes (Fig. 1.26). None of its permanent inhabitants is under 65 years of age! It is now a *very* quiet place.

One of the island's two primary schools has closed down recently; older pupils have to live in Oban during term-time to attend the secondary school there. Lismore's only general store could not survive without the extra trade which the tourists bring. The post office closed for a few years because it was not possible to replace the postmaster, who had died in 1970.

▲ **Figure 1.22** Early migration was far from peaceful

▼ **Figure 1.23** Lismore Island in the Inner Hebrides

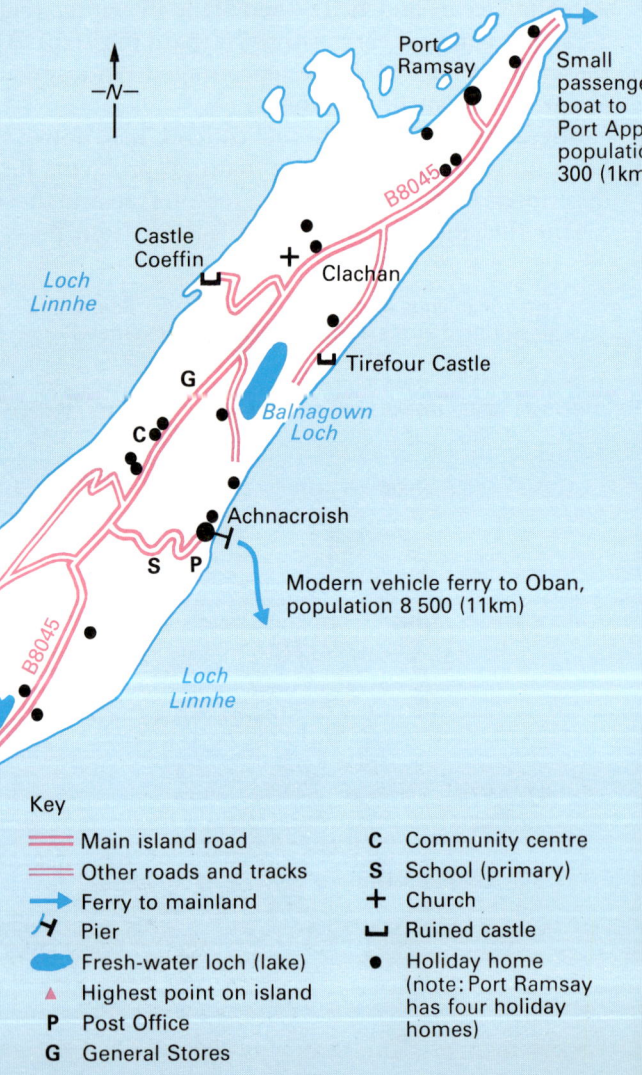

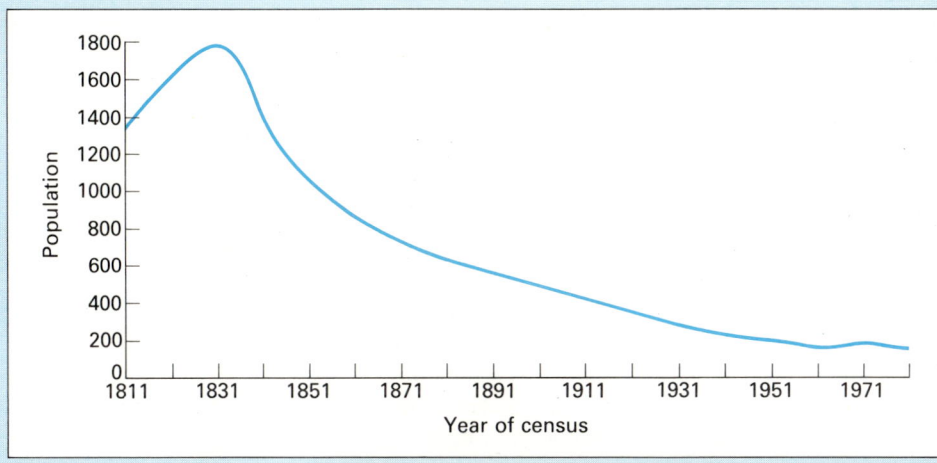

▲ **Figure 1.24** Line graph showing changes in Lismore's population, 1811–1981

▼ **Figure 1.25** Abandoned croft on Lismore

► **Figure 1.26** The main street of Port Ramsay, Lismore

The continuing decline of population in northern Scotland has caused great concern. In 1965, the government created the Highlands and Islands Development Board to encourage new ideas likely to create employment. It has **subsidised** the building of hotels and fishing boats; many road and ferry links have been improved with money received from the Board.

Some islands in the region have been much more fortunate than Lismore. The Orkneys and the Shetlands (Fig. 1.20) in the far north are near to oilfields in the North Sea and the oil industry has created hundreds of well-paid jobs in both groups of islands. Their populations have increased quite sharply as a result. In the mid-1980s, however, the price of crude oil slumped and this created many unexpected problems for these islands. Unemployment suddenly increased and it remains to be seen whether their populations will begin to decline once again.

Urbanisation can also be a mixed blessing for rural areas in *developing* countries. Although its effects are very similar, its causes are somewhat different. In the developing countries it is mainly poverty which drives people into the cities.

The loss of young people is keenly felt by rural areas throughout the world, and Question 3 e asks you to think of reasons why this should be so. These people can cope best with the hard physical work demanded by traditional Third World farming methods. Some migrants like to keep parcels of land in the country-side, and return to their villages for a few months each year to add to their food supply. This is usually done during the rainy season, when the soil is softer and the crops grow best.

Ambitious country-dwellers move because they can't 'get on' on rural areas. A particularly serious loss is the migration of skilled craftsmen such as tailors, potters and metal workers – all of them key members of isolated rural communities. They cannot compete with the much cheaper, mass-produced goods made in the cities, and so go out of business.

Many developing countries are now trying much harder to boost the economies of their rural areas, even though the resources they have to do this are quite limited. For example, the Indian government is subsidising many small village spinning and weaving businesses. Ghana is making it possible for villages to have a new school or build an extra classroom. It is doing this as part of a nationwide 'self-help' campaign; the government provides free cement and expert guidance, and the villagers do the actual building in their spare time. Both types of action should tempt the young and able people to stay in rural areas. Investment in new farming methods helps to make agriculture a more efficient, and hence more attractive occupation. Unfortunately, increased efficiency usually means fewer jobs – and this *encourages* migration!

1 How can urbanisation help overpopulated rural areas?

2 List separately the pull and push factors which have encouraged migration from the Highlands and Islands region of Scotland over the last 250 years.

3 Copy out and complete these statements:
a The ferry journey from Oban to Achnacroish Pier on Lismore Island is only . . . km long.
b A secondary school has not been built on Lismore because . . .
c It took a long time to replace Lismore's post-master because . . .
d Lismore's shop keeper has to keep a stock of goods which are popular with tourists because . . .
e The migration of young people from Lismore is particularly serious because . . . (Try to think of your own reasons for e and f.)
f The islanders do not like having so many holiday homes because . . .

4 Imagine that you live on Lismore.
You fear that de-population is now threatening the very existence of your island community and feel that you must do something about the situation before it is too late. After discussing in class the various measures which might be taken to reduce migration to the mainland, write a 'strong' letter to the local Member of Parliament which:

□ points out the effects which migration has already had on your island.
□ states your fears for the future.
□ suggests new measures to maintain and perhaps increase the island's population in the future (e.g. involve the Highlands and Islands Development Board).

5 **a** Study the population census figures in the table below; Stornoway is the main town of the Western Isles county (see Fig. 1.20).
b Summarise the population trends shown in the table for:

□ the whole of Scotland
□ the Western Isles county
□ the town of Stornoway
□ the *rural* areas of the Western Isles.

6 **a** How are some Third World countries trying to discourage rural to urban migration?
b What other measures do *you* think they should take to encourage young adults to stay in their villages?

Population	in 1921	in 1951	in 1961	in 1971	in 1981
Scotland (x 1000)	4882	5096	5178	5229	5229
Western Isles	44177	35591	32609	30467	31879
Stornoway	4100	5000	5400	5300	8700

Urbanisation in Britain

2.1 Introduction: nineteenth-century urbanisation

Before its Industrial Revolution (about 1760 to 1850), Britain's economy was based mainly on agriculture. Most people lived in small villages by the coast or on fertile lowlands (Fig. 2.1), and the towns were much smaller than they are today. These settlements had survived for many centuries because they were built on good **sites** (places). The most popular types of site were:

□ on river estuaries;
□ in sheltered bays;
□ at 'bridging points', where rivers were narrow enough to build bridges across them;
□ at 'confluences', places where rivers are joined
□ by springs, which provide a constant supply of drinking water;
□ by fast-flowing rivers able to turn the water-wheel of a mill;
□ in valleys, where it was easier to build roads;
□ at the top of steep-sided hills, which could be defended against attack.

The changes in **population distribution** brought about by the Industrial Revolution can be seen in Fig. 2.2. By 1850, the British economy had turned to mining, manufacturing and transport, and fewer people were engaged in agriculture. Large numbers of families had left the rural areas to work in the towns, and this put great pressure on their stock of housing. An added problem was population *growth*. In 1750, Britain's population was only 8 million; by 1851 it had climbed to 21 million – a rise of 163%. Most of this increase was absorbed by the towns. The combined effect of rural to urban migration and population increase can be seen from these figures for Preston. In 1801 this busy cotton-weaving town in central Lancashire had 11 887 people. Only 20 years later, this figure had more than doubled – to 24 575.

Builders in England and Wales responded at first by erecting small dwellings called **back-to-backs** (Figs 2.3 and 2.4). Few examples of this type of terraced housing can be seen today because the back-to-backs were cramped and made of the cheapest materials available.

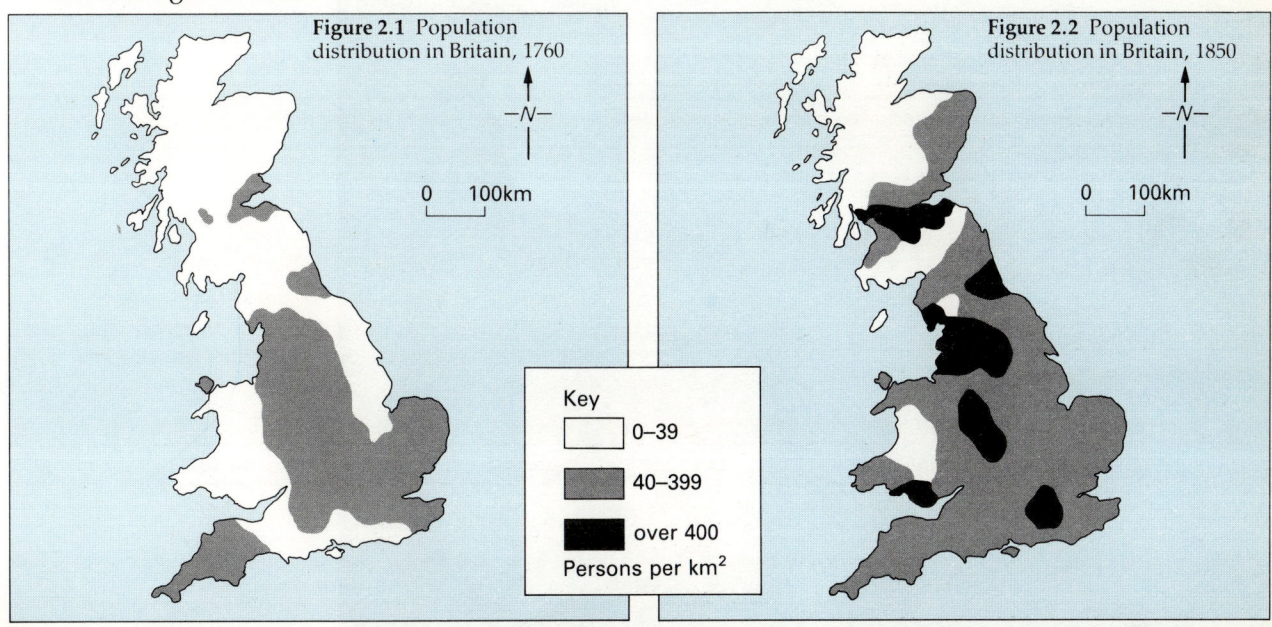

Figure 2.1 Population distribution in Britain, 1760

Figure 2.2 Population distribution in Britain, 1850

Key
0–39
40–399
over 400
Persons per km²

▲**Figure 2.3** Layout of back-to-back housing in Burnley

◄**Figure 2.4** Back-to-back housing in Leeds

◄**Figure 2.5** Glasgow tenements viewed from the back in 1945

The name 'back-to-back' was very appropriate as an inner wall separated the houses at the front and rear of the terrace. The door of each house led directly from the street into a single living room on the ground floor, and a narrow staircase led to the sleeping area upstairs. Some of them had a cellar. This could be a dangerous place as many back-to-backs were built on marshy land. This often flooded after heavy rain and a number of people in Liverpool were trapped in the cellars of their back-to-backs, and drowned.

You might think that the houses shown in Figs 2.3 and 2.4 would provide enough space for a working-class family of the time – but each house fitted on to a piece of ground only 4 m^2 – the size of a *single* room in a typical modern house. The problem of space was made much worse because the builders couldn't meet the demand for new houses. So individual rooms were sub-let (rented out) to other families, with some quite horrifying results. One serious case of over-crowding was discovered in Liverpool, where no fewer than 68 workers were found to be living in one small terraced house. A survey carried out in 1842 in Bury (a cotton-*spinning* town) revealed that 800 families were sleeping three to a bed, 200 four to a bed, and 63 families *five or more* to a bed!

Poor health was another major problem. The back-to-back houses were poorly ventilated (lacked fresh air) because their dividing walls prevented through-drafts which could dry them out. They were also built without damp-courses in the walls. Dampness would rise from the ground, up the walls and eventually rot the whole building. Tuberculosis – a disease of the lungs – was very common under such conditions. Smoke from nearby mill and factory chimneys blackened the houses around them and ruined the health of their occupants. The streets were rarely surfaced properly and became dumping grounds for all kinds of household rubbish – ideal breeding conditions for rats and flies. At best, the people looked pale, tired and depressed. Groups of back-to-backs had to share a single 'privvy' (outside toilet), water tap and wash-house. There were frequent outbreaks of cholera and typhoid due to the insanitary conditions. It took a series of *very* strict Public Health Acts to improve the standards of hygiene in British late nineteenth-century towns.

Scotland also suffered a serious lack of housing, but the builders there had a different way of cramming homes onto the land. They built **tenements** (Fig. 2.5) – blocks of flats usually four storeys high. These were arranged around a central courtyard in which communal (shared) privvies, water taps and wash-houses were sited. One such area was the Gorbals, a district near the heart of Glasgow, where the inhabitants had to use packs of dogs to chase the rats away. The tenements made Glasgow the most overcrowded city in Europe. Poor health, drunkenness and a high crime rate were the result.

Housing standards improved considerably after 1875, when an Act of Parliament laid down minimum building standards for the future making it illegal to erect houses which did not match up to these new, higher, standards. The dwellings built after 1875 were called **by-law houses**. This was because the government told every town council to pass its own local rules – called 'by-laws' – to make sure the Act was obeyed. This type of housing was a great improvement on the back-to-backs which it replaced.

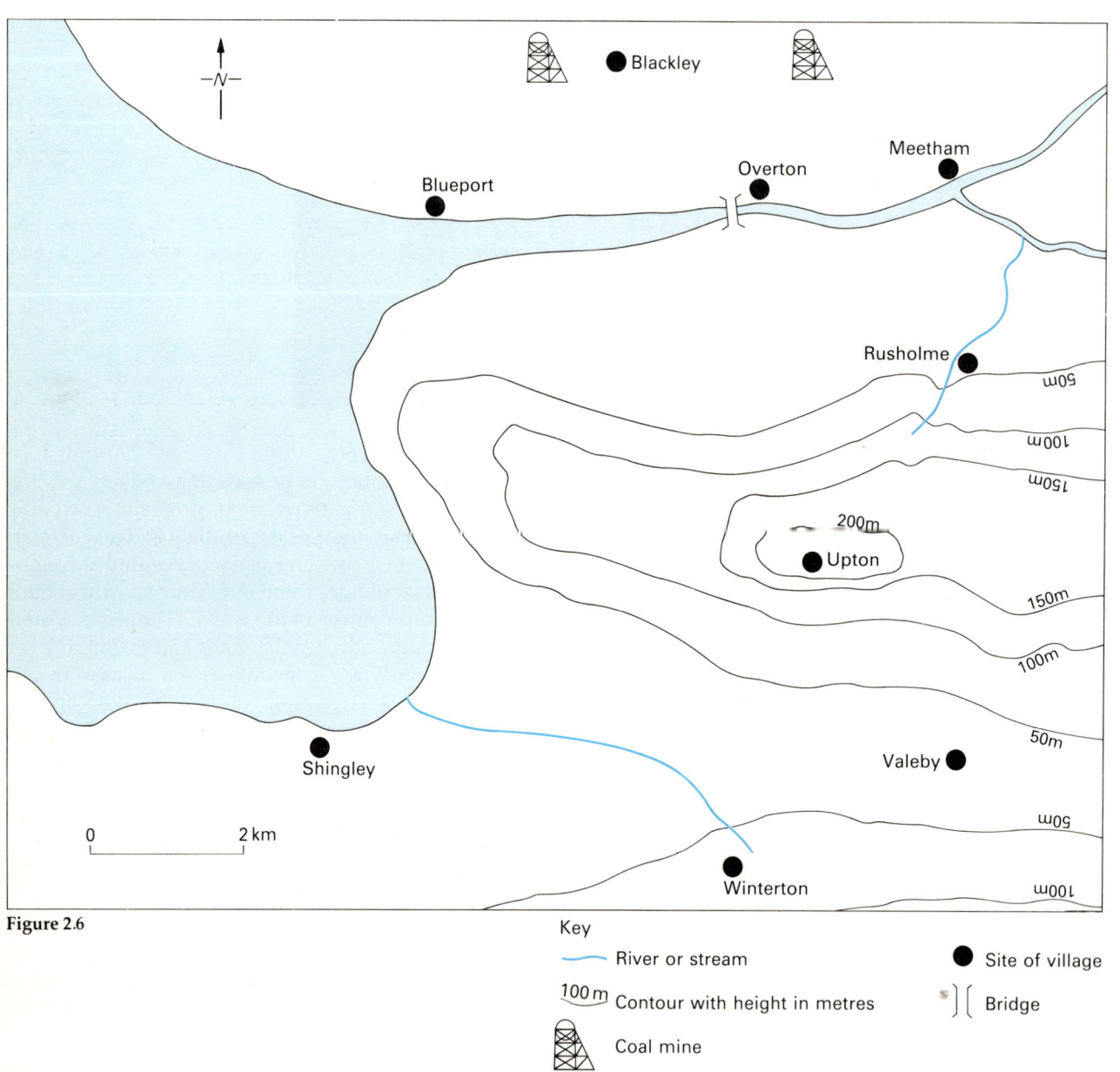

Figure 2.6

Key

— River or stream

100m Contour with height in metres

Coal mine

● Site of village

) (Bridge

Towards the end of the nineteenth century, the more highly skilled workers could afford something a little better than the standard by-law type of house. They wanted a small garden and, at long last, traces of the countryside began to appear in our towns and cities.

1 a Pair up the names of the villages on Fig. 2.6 (except Blackley) with the types of site listed on page 21.
b Explain why Blackley and Blueport were *most* likely to have grown rapidly during the Industrial Revolution.

2 Pair up the names of these villages with the descriptions of their sites (see the Ordnance Survey map extract on page 148):

Leybourne	at an important road junction
Ryarsh	by a castle
Snodland	in a valley and by a stream
Wrotham Health	on the banks of a wide river
Vigo Village	on the crest (top) of a steep-sided ridge.

3 a Describe the distribution of Britain's population in 1760.
b How was the distribution of Britain's population in 1851 different to that in 1760?
c After class discussion, suggest reasons for these changes in population distribution.

4 You have been asked to write a scene for a play about living conditions on nineteenth-century British industrial towns. The Director of the play has given you this information:

'The scene is the courtyard in the middle of a block of back-to-back houses. There is a water pump in one corner of the yard, and near to it is a row of three privvies. Washing, rubbish and people seem to be everywhere. It is a hot day, and the smell of rotting refuse is sickening. The children do not have proper shoes and their clothes are in tatters. Their parents are not very old but they look pale and worn out with working long hours in the mill nearby. One man is lying on the floor. He was paid yesterday in the pub at the corner of the block and spent all his money on drink. His wife and children have thrown him out of the house, disgusted because they may starve next week.'

There are two speaking characters in the scene:

THE MILL OWNER He has had the houses built for his workers and also owns the pub at the corner. He is proud of giving work to all these people. He is also pleased at having provided houses for his workers, because otherwise they would be sleeping on the streets like many others who have just arrived in 'Newtown', a typical cotton textile town in Lancashire. He believes it is the *people* who have turned his houses into slums.

MRS PEABODY She is also wealthy but has dedicated her life to helping the poor. She is keen to build some housing for her own workers, but doesn't want to repeat the mistakes made by the greedy mill owners. She thinks the state of the courtyard, the houses and the people are the fault of the Mill Owner, the local town council and the government. But she does *not* blame the workers.

Your task is to write about two pages of discussion between the two characters. They will argue, because one of them has sympathy with the workers and the other does not. This is how your scene *could* begin, but you may choose a different opening if you wish.

MRS PEABODY: Good morning, Mr ... It is good of you to meet me to discuss this block of houses you built five years ago.
MR ...: Why should I not wish to meet you? I know you think that I am responsible for the misery of my workers, but I am proud of what I have done.
MRS PEABODY: I cannot possibly agree with you. Now look here at ...

5 *Briefly* describe the 'tenement' system of housing used in Scotland.

6 a Why were by-law houses given this name?
b Do you believe the government raised housing standards willingly, or because they were under great pressure to do this?
c Suggest reasons why most by-law houses were built *only* to the minimum standards laid down by the government.

2.2 *Urban zones*

Every town is different, and there are many reasons why this should be so. Site (see page 21) may limit the directions in which a settlement can grow and it can also restrict its size. For example, a town built in a narrow valley will tend to become **linear** – long but narrow in shape. Many coastal holiday resorts are linear as their main business areas have to be near to the sea. This is because visitors wish to stay and relax as near to the beach as possible (Fig. 2.7). *Residents* are more concerned by the distance they have to travel to work and to shop within the town centre. Because of this, most towns tend to be **nucleated** – much more rounded in shape. Manchester is a well-known example of this second type of urban growth.

The *inner* layout of a town can often tell us a great deal about its history. Some towns are very old indeed, and their basic street patterns may not have changed for hundreds of years. Others expanded rapidly during the Industrial Revolution and so have many streets grouped together in a regular 'grid' pattern. Some are so modern that their layouts are quite different to that of any of the older types of settlement. It may therefore surprise you to know that *most* of our towns are quite similar in many ways.

The easiest way to study the layout of any town is to look for its **urban zones**. A zone is an area which serves a particular purpose, such as housing, and is unlike the other areas around it. Figure 2.8 shows the pattern of zones in a typical nucleated British town. The rest of this unit will help you to recognise different types of zone in the towns in your local area.

The Central Business District

The **Central Business District**, often shortened to CBD, is the zone at the heart of a town, (Fig. 2.9). It may be very old, which explains why its streets do not have any regular pattern (Fig. 2.10). The CBD provides most of the services needed by a town, and

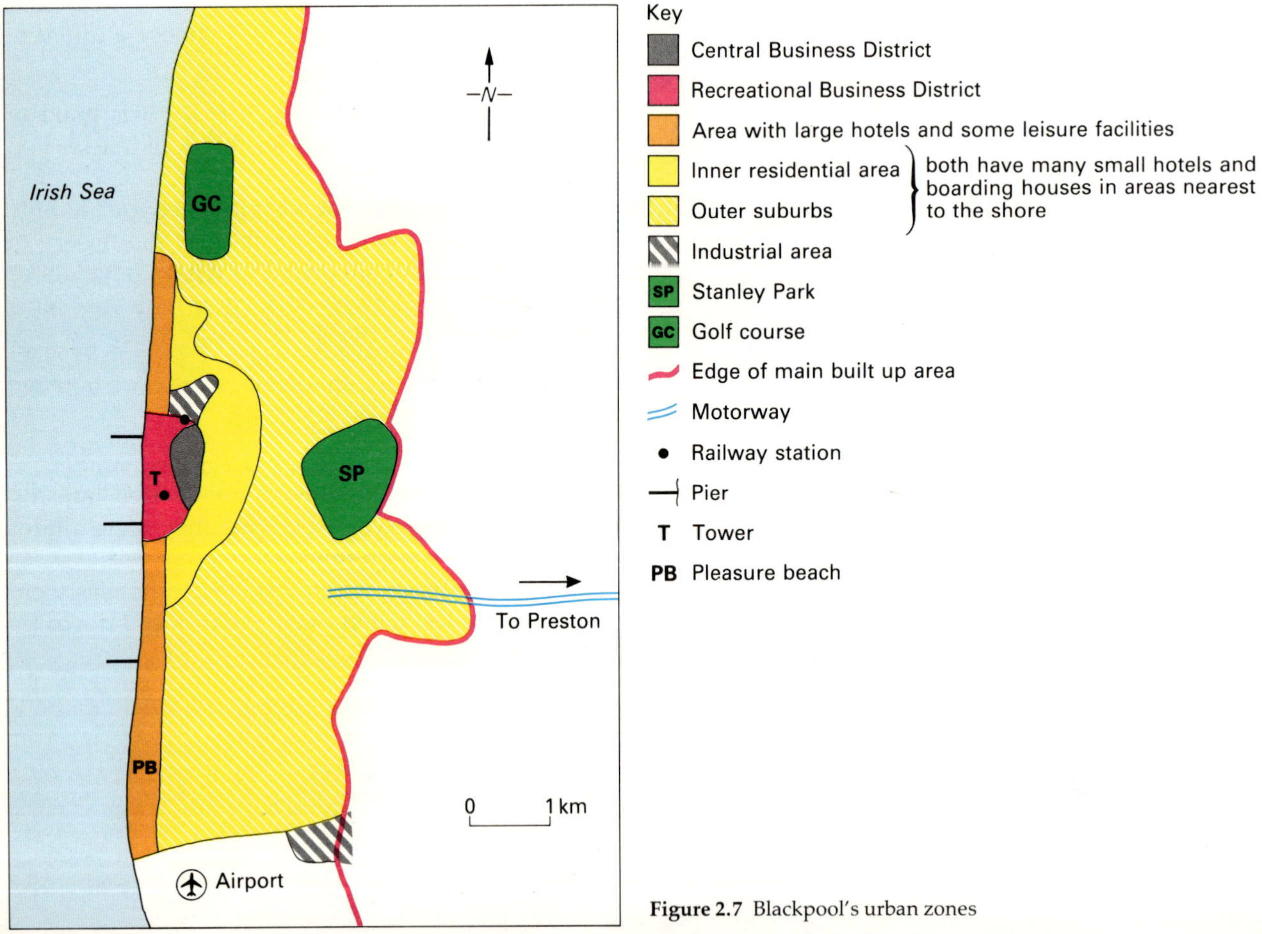
Irish Sea
GC
SP
T
PB
N
To Preston
0
1 km
Airport
Key
Central Business District
Recreational Business District
Area with large hotels and some leisure facilities
Inner residential area
Outer suburbs
both have many small hotels and boarding houses in areas nearest to the shore
Industrial area
SP Stanley Park
GC Golf course
Edge of main built up area
Motorway
Railway station
Pier
T Tower
PB Pleasure beach

Figure 2.7 Blackpool's urban zones

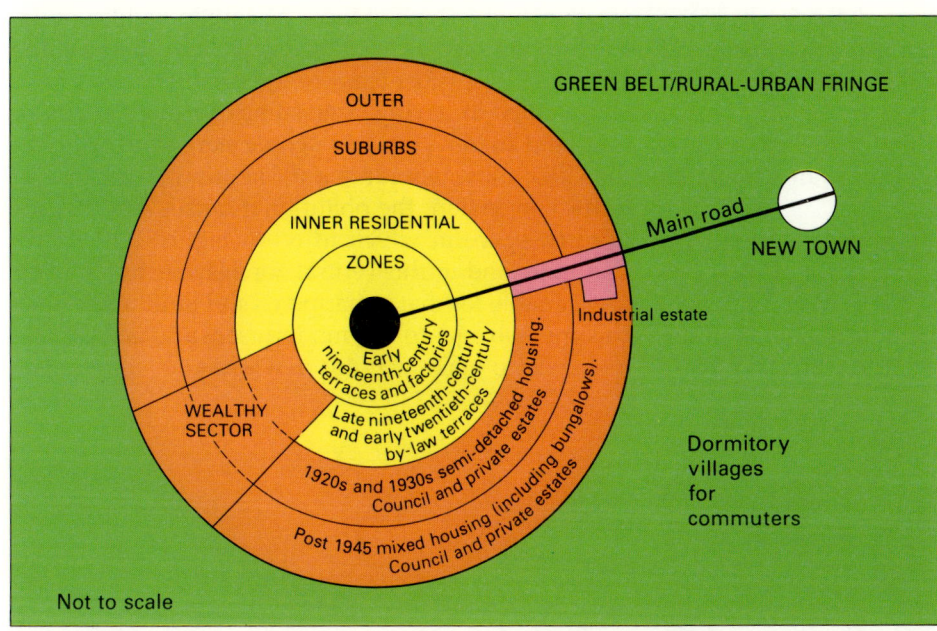

Key
- ■ Central Business District
- ■ High quality housing
- ■ Lower quality housing
- ■ Industrial zone
- ■ Open countryside, with villages

GREEN BELT/RURAL-URBAN FRINGE

OUTER

SUBURBS

INNER RESIDENTIAL

ZONES

Early nineteenth-century terraces and factories

Late nineteenth-century and early twentieth-century by-law terraces

1920s and 1930s semi-detached housing. Council and private estates

Post 1945 mixed housing (including bungalows). Council and private estates

WEALTHY SECTOR

Main road

NEW TOWN

Industrial estate

Dormitory villages for commuters

Not to scale

◀ **Figure 2.8** Zone layout of a typical British city

▼ **Figure 2.9** An aerial view of Manchester's Central Business District

many people work there during the day. It is the administrative centre, based on the town hall, and has the largest 'public buildings' (e.g. the main library). The chief transport depots are usually located inside or on the edge of the CBD. If the town has a cathedral or ancient university, it too can usually be found in this central zone.

As its name suggests, the CBD is the most common place to build banks, shops and blocks of offices. The largest stores are there because it is so easy for people to travel into the CBD from the housing areas in the town and the villages around it. There are also many smaller shops, some of them dealing in highly specialised goods such as expensive jewellery. Because the CBD is so accessible (easy to get to), businesses have felt justified in paying large sums of money to buy or rent land there. Very few people can afford to *live* in the CBD; most wouldn't want to anyway because of traffic congestion, the pollution it causes and the lack of open space in which to relax. Figure 2.10 shows that little land in this central zone is vacant – it is far too valuable for that to happen – and buildings which become too old to be modernised are quickly demolished to make way for new ones.

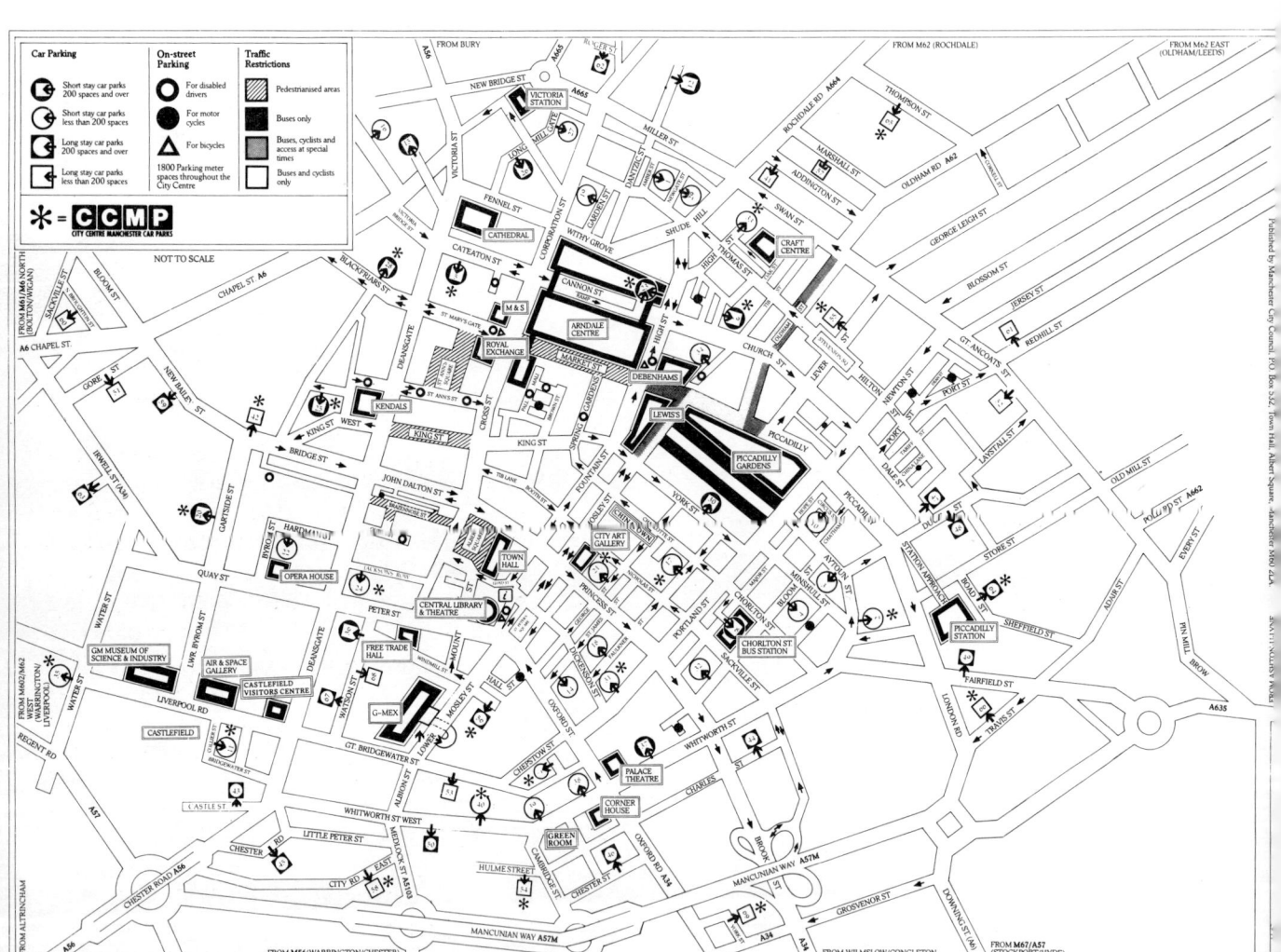

Figure 2.10 Street plan of Manchester's CBD

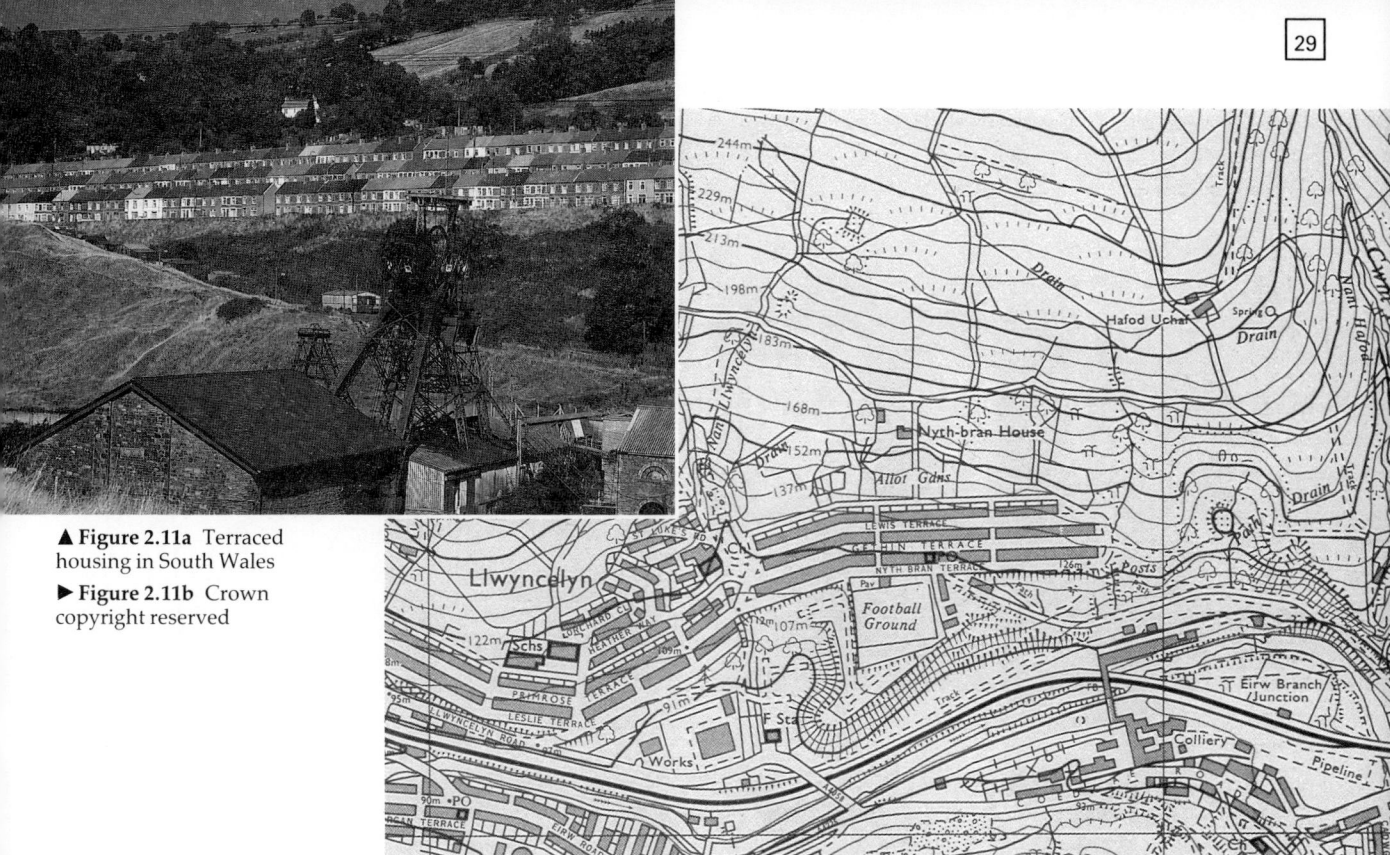

▲ **Figure 2.11a** Terraced housing in South Wales

▶ **Figure 2.11b** Crown copyright reserved

The inner residential zone

This second type of zone dates back to the nineteenth century and the early years of the twentieth. It consists of the type of terraced housing described in the last unit, and so the best examples are to be seen in those towns which grew very rapidly during the Industrial Revolution. Figure 2.11 shows the general layout and appearance of an inner residential zone in a narrow valley in Wales.

The older parts of these zones are now being re-built, and this process is described in units 2.5 and 2.6. These areas are sometimes called **twilight zones** or **shatter belts** because of the great changes which are taking place within them. The inner residential areas offer moderately priced housing which is very attractive to families buying their first house, as well as immigrants who have only recently entered this country.

The outer suburbs

Our towns expanded even more rapidly during the 1920s and 1930s – the years between the two World Wars. This growth was encouraged partly by improvements in public transport (e.g. electrically-powered trams, diesel-engine buses and faster local train services). It was also due to a great increase in the number of privately-owned cars. These changes allowed **commuters** to live much further from their place of work.

By that time, working people had become tired of living in a depressing maze of houses, mills, canals and railway lines. They felt a great need to have a garden of their own and live within easy reach of open countryside. The result was the building of large **suburbs** (areas of housing on the edge of a town). The two-storey, semi-detached type of house shown in Fig. 2.12a was extremely popular at that time. House-buyers preferred it because it offered far more privacy than the nineteenth-century terraces and gave them enough room for a garden and a garage. The builders chose it because it used less land and was cheaper to build than the detached type of house. This highly successful design was also widely used on council estates of rented accommodation.

The first 'semis' were built on or near the main bus routes, as these provided easy access into the town centres. This kind of urban growth is called **ribbon development**. When all the main roads had been built up as far as people were willing to travel to work, the builders moved onto the open fields between them. This second stage of building is called **in-filling**.

▲ **Figure 2.12a** 1930s semi-detached houses, Hatfield, Hertfordshire

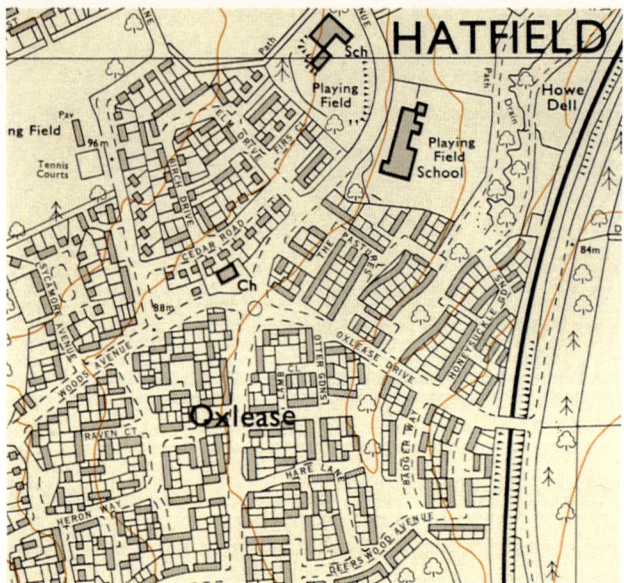

▲ **Figure 2.12b**

▲ **Figure 2.13a** Post-war bungalow, Blaydon, Tyne and Wear

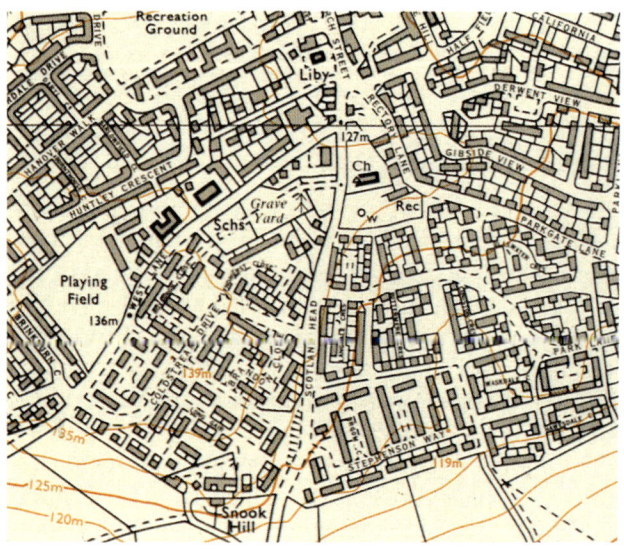

▲ **Figure 2.13b** Crown copyright reserved

The builders took care to space out their new houses and to plant trees wherever possible. They built much wider roads to take more traffic and make street-parking easier. They tried to avoid boring, re-petitive road patterns by including crescents (curved roads) and cul-de-sacs (dead-end-roads). The system of naming roads changed too; 'streets' and 'roads' became unfashionable, but 'avenues', 'crescents', 'lanes', 'groves' and 'closes' were definitely in! Schools built during this period had much larger playing areas, which increased the amount of open space within the suburbs.

Urban sprawl continued when Britain began to recover from the Second World War. Very few houses had been build during the six years of fighting and so there was a great demand for new dwellings to buy and council houses to rent. Large numbers of the ever-popular semi continued to be built, but bunga-lows provided a welcome alternative for older people (Fig. 2.13). About this time, town councils gained control over planning matters. This meant that they could decide where and how future building deve-lopments would take place. The units on new towns describe some of the ways in which the planners used their new powers.

1 a Explain what shapes 'linear' and 'nucleated' settlements have.
b Why does Blackpool have a linear shape?
c Why do *most* settlements have a nucleated shape?

2 With the help of an Ordnance Survey map of your local area:
a Name at least one linear and one nucleated town or village.
b After class discussion, suggest reasons why your chosen settlements have these shapes. (Hint: consider relief, rivers, the coastline and roads.)

3 Draw a transect (a straight-line cross-section) northwards from the centre of the town shown in Fig. 2.8. Your transect must show the boundary lines between the four zones, which need to be labelled with their names.

4 a Name the type of urban zone which Blackpool has, but the imaginery town in Fig. 2.8 does not.
b What other differences do you notice between the zone-layouts shown in Figs 2.7 and 2.8?

5 Pair these important terms with their correct meanings:

Central Business District (CBD)	an area of housing on the edge of a town
In-filling	
Ribbon development	a part of a town which is different to the other areas around it
Suburb	building along a main road
Urban zone	building between main roads
	the heart of a town where many people work, but few actually live

6 Copy out this table, then complete it with the help of the text and Figs. 2.11, 2.12 and 2.13.

	Inner residential zone	Inner suburbs	Outer suburbs
Main building period			
Description of street pattern			
Main features of housing			
Any other differences			

7 a List the different Central Business District land uses shown in Fig. 2.10 – under these three headings.
□ public buildings
□ transport
□ commerce (i.e. business).

b What do you notice about the street layout of this area?

c Try to suggest reasons for your answer to b above.

8 Explain why many of our towns expanded very quickly:
a in the nineteenth century
b between 1918 and 1939
c after 1945

9 After class discussion, state how the layout of your own/a nearby town compares with the 'typical' layout shown in Fig. 2.8. You will need to draw (or borrow!) an urban-zone map of your chosen town to do this.

2.3 Residential zones in Leyland

Leyland, in Lancashire, has many of the types of residential zone described in the previous unit. Although world-famous for building lorries and buses, Leyland has concentrated on farming for most of its long history. The town's agricultural past is recalled today in street names such as Fox Lane, Broadfield Drive and Golden Hill Lane (named after the corn which once grew there). Fox Lane lies close to The Cross, the ancient heart of Leyland. The chief shopping streets of twentieth century Leyland strike northwards from this point.

Industrialisation came late to Leyland, mainly because much of the area was owned by two wealthy families who didn't like parting with their land! Leyland was linked to the national railway network in 1838. This event led to the building of a few small cotton-weaving mills on the western side of the town, as well as some terraced housing characteristic (typical) of that period. Both developments were, however, on a much smaller scale than in the nearby cotton textile towns of Blackburn and Preston.

In 1895, the Lancashire Steam Motor Company was founded, and started business on a back street near to the town centre. It began by building lawn mowers and road vehicles powered by steam, then expanded more rapidly after developing a reliable petrol-driven engine. The works later moved to a much larger site on the northern edge of the town and attracted many newcomers into Leyland. Much of the town's housing stock is of the semi-detached type erected during the 1920s and 1930s, when the motor works expanded yet again. The town council has also built a number of estates of rented accommodation.

The most recent phase in the town's development began in 1971, when it was announced that Leyland, Preston and Chorley were to be the key centres for the Central Lancashire New Town (Fig. 2.14). This led to the building of many new houses and factories, mainly in the southern and western districts. It also meant that Leyland's population has increased – at a time when *most* of our industrial towns are losing people.

Figure 2.15 shows the present layout of the town. It includes the positions of five quite different residential zones which are examined in the rest of this unit. The information on these zones has been taken from the 1981 national population census. For census purposes, every British town is divided up into **wards** and **enumeration districts**. The wards are quite large areas with 2 000 – 8 000 people; Leyland has six, most

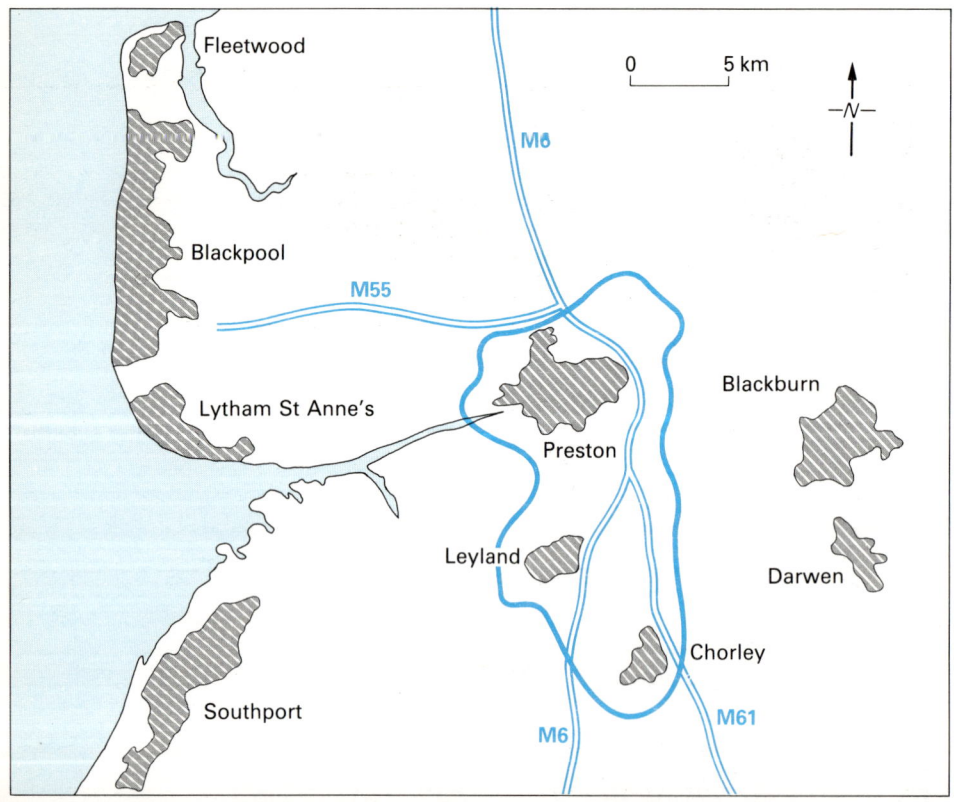

◀ **Figure 2.14** Central Lancashire New Town

Key

▨ Existing built-up area
Boundary of Central Lancashire New Town
Motorway

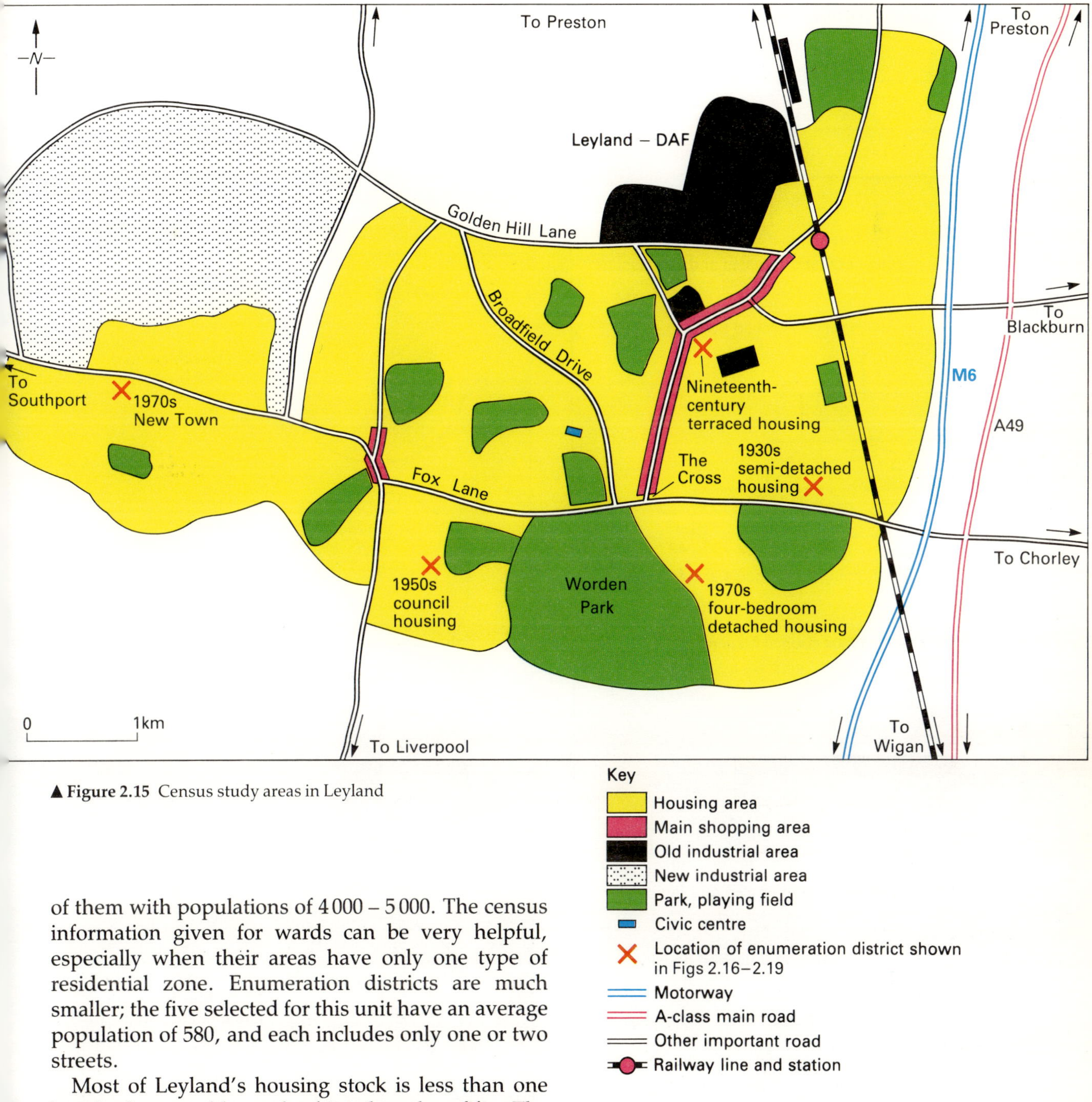

▲ **Figure 2.15** Census study areas in Leyland

Key

- Housing area
- Main shopping area
- Old industrial area
- New industrial area
- Park, playing field
- Civic centre
- ✕ Location of enumeration district shown in Figs 2.16–2.19
- Motorway
- A-class main road
- Other important road
- Railway line and station

of them with populations of 4 000 – 5 000. The census information given for wards can be very helpful, especially when their areas have only one type of residential zone. Enumeration districts are much smaller; the five selected for this unit have an average population of 580, and each includes only one or two streets.

Most of Leyland's housing stock is less than one hundred years old, much of it is less than fifty. The choice of enumeration districts reflects this, as only one of them was completed before 1900. The remaining illustrations provide four types of information about these districts. Information has not been provided on the proportions of 'sub-standard' housing (e.g. without an inside bathroom or toilet) or the 'ethnic origins' of the population, as both figures for Leyland are extremely low.

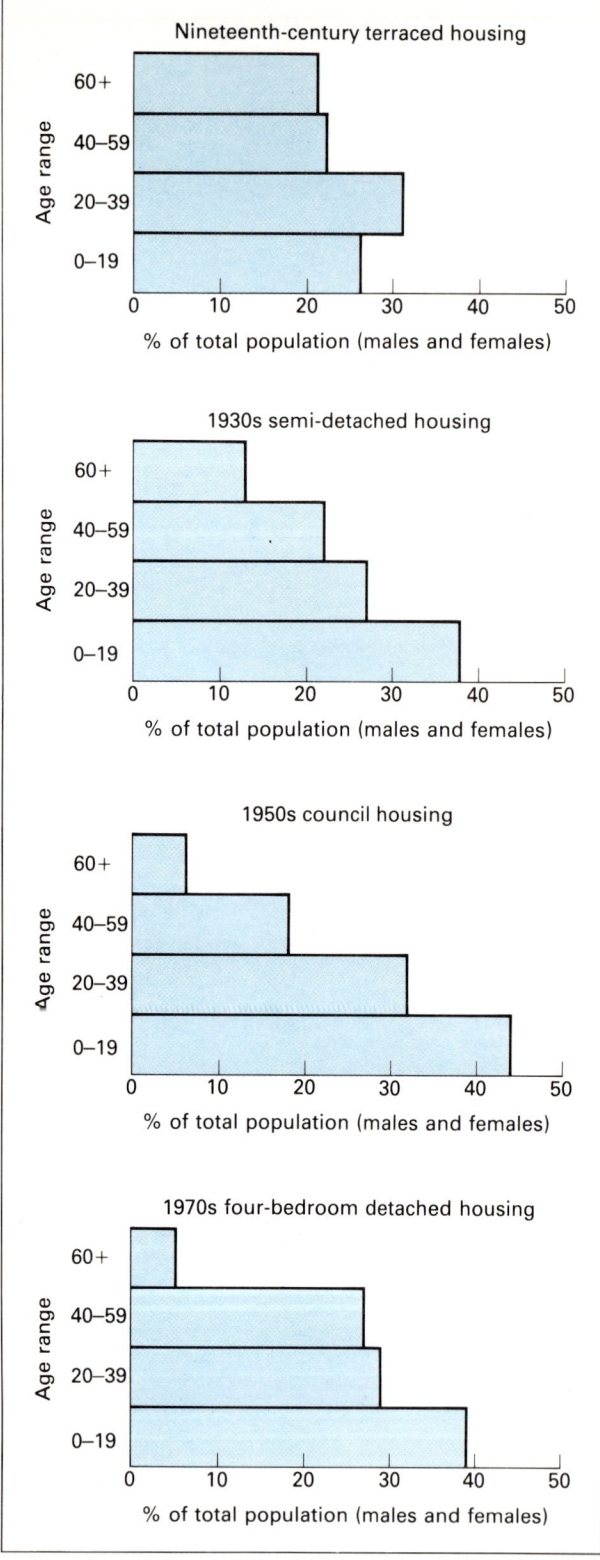

Figure 2.16 Population pyramids for four Leyland enumeration districts

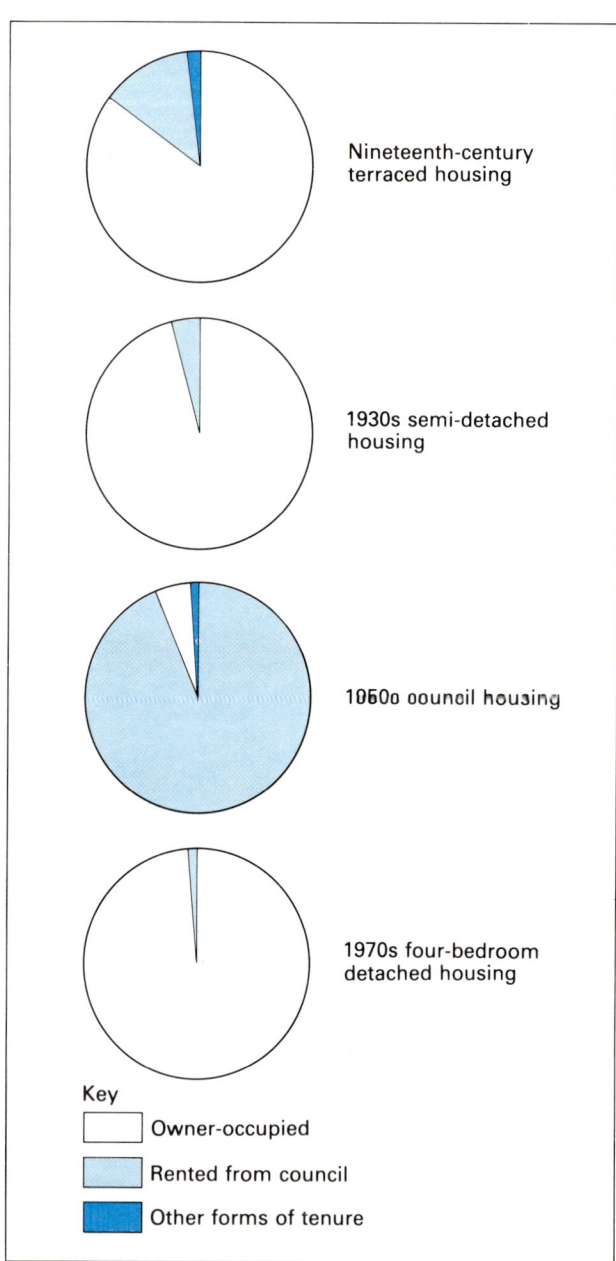

1 a What are the differences between a ward and an enumeration district?
b Why is it usually better to use the census information given for *enumeration districts* when comparing residential zones?

2 Explain what is meant by the terms:
house tenure, owner-occupied, population pyramid.

Figure 2.17 Pie graphs showing house tenure in these enumeration districts. 'Tenure' means kind of ownership

3 Draw the illustrations in Figs 2.16–2.19 then complete them using this information for the '1970s New Town' district:

a population age structure (population pyramids)

0–19 years	34%
20–39 years	44%
40–59 years	14%
60+ years	8%

b house tenure (pie graphs)

Owner-occupied	22°
Rented from council	324°
Others	14°

c *average* car ownership (bar graph)
0.85 cars per household

d car ownership (divided bar graph)

0 cars	28%
1 car	61%
2 cars	9%
3 or more cars	2%

4 Which districts have:
a The highest % of young people aged 0–19 years?
b The highest % of young adults aged 20–39?
c The highest % of retired people?
d The *two* highest percentages of owner-occupied housing?
e The *two* lowest percentages of owner-occupied housing?
f The *two* highest figures for *average* car ownership?

g The lowest figure for *average* car ownership?
h The highest proportion of households without a car?
i The highest proportion of households with *one or more* cars?
j The highest proportion of households with *two or more* cars?

5 Which districts appear to have:
a The most 'balanced' population (i.e. with similar numbers of people in each of the four age ranges)?
b The most balanced pattern of house tenure?
c The wealthiest group of people?
d The least wealthy group of people?

6 After class discussion:
a Suggest ways in which these paired types of information appear to be linked.

☐ *population age structure* and *house tenure*.
☐ *population age structure* and *car ownership*.
☐ *house tenure* and *car ownership*.

b Suggest *reasons* for any links you have identified in **a**.

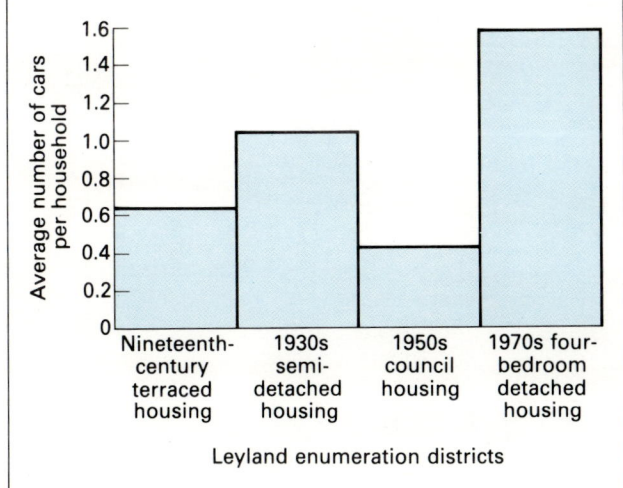

▲ **Figure 2.18** Bar graph showing the average car ownership in these Leyland enumeration districts

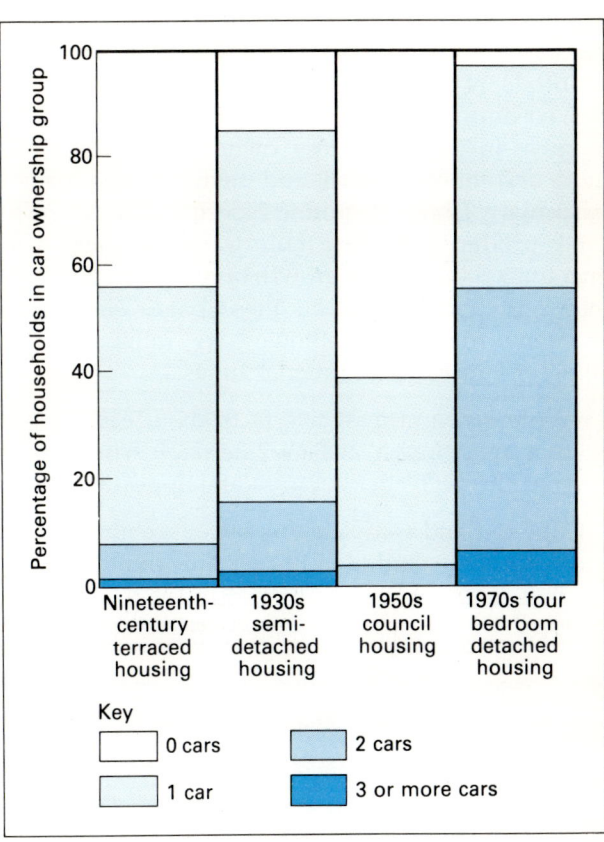

▲ **Figure 2.19** Multiple divided bar graph showing car ownership patterns in these Leyland enumeration districts

2.4 Industrial zones

Industry uses less land than housing, but it too can be found in almost any part of a large town. Until the 1800s, it was usual to build factories and mills on river banks; road surfaces were generally poor at that time and so it was much easier to carry bulky, heavy materials such as coal, by water. The end of the eighteenth century saw the first great improvement in transport. 'Canal mania' was at its height, and every sizeable town clamoured to have one of the new man-made rivers routed through it. The canal era was quite short-lived, however, for a much more efficient form of transport appeared towards the middle of the next century. By the 1870s, *every* industrial town in the country had been linked to the railway network. Many factories wanted to have their own sidings to unload raw materials and despatch finished goods and they could only do this by being sited near to a main line.

Transport patterns began to change yet again after the First World War (1914–18) and so of course did those of factory sites. The war had quickened research into the design of commercial vehicles and their engines, and by the 1920s British factories could produce thousands of lorries every month. The country's network of main roads was greatly improved during the 1920s and 1930s to cope with the increase in traffic. Factory owners used the roads more and more for small and medium sized loads – particularly those requiring speedy delivery. The 1960s produced our first motorways and these have encouraged the further growth of road transport. The map and photographs on these pages show how changing transport methods have influenced the siting of Lancaster's industrial zones.

1   Write down at least five facts about each industrial area shown in Figs 2.20–2.24. You should consider both their location and appearance.

2   Copy out and complete this table (see jumbled list of words for writing in the second column).

Period	Latest form of transport
Before 1790s	
1790s–1830s	
1840s–1910s	
1920s–1950s	
1960s onwards	

List: canals, motorways, railways, rivers, roads.

▲ **Figure 2.20** Industrial estate between Morecombe and Lancaster

▲ **Figure 2.21** Riverside industry west of Lancaster city centre

▲ **Figure 2.22** Roadside industry near to junction 34 on the A683

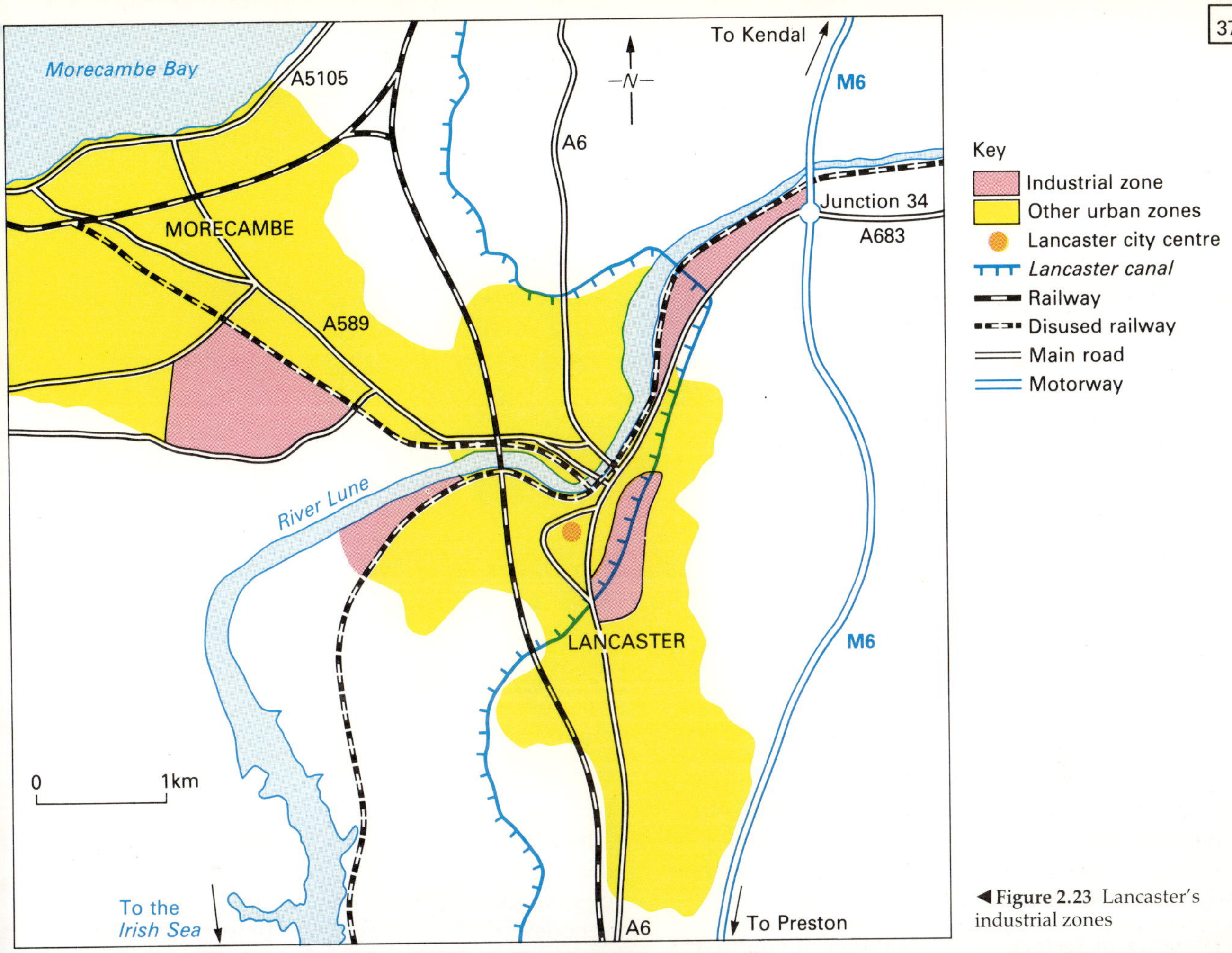

Key

▨	Industrial zone
▨	Other urban zones
●	Lancaster city centre
⊞⊞⊞	*Lancaster canal*
▬▬	Railway
▬▬	Disused railway
══	Main road
══	Motorway

◄ **Figure 2.23** Lancaster's industrial zones

3 a List in rough any advantages which you consider:

- ☐ canals have over rivers
- ☐ railways have over canals
- ☐ roads (generally) have over railways

for carrying freight, *not* passengers.

b Discuss your lists openly in class.

c Write a paragraph of 'decent English' on each comparison – based on your shared ideas.

▲ **Figure 2.24** Old textile mill east of Lancaster city centre

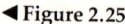

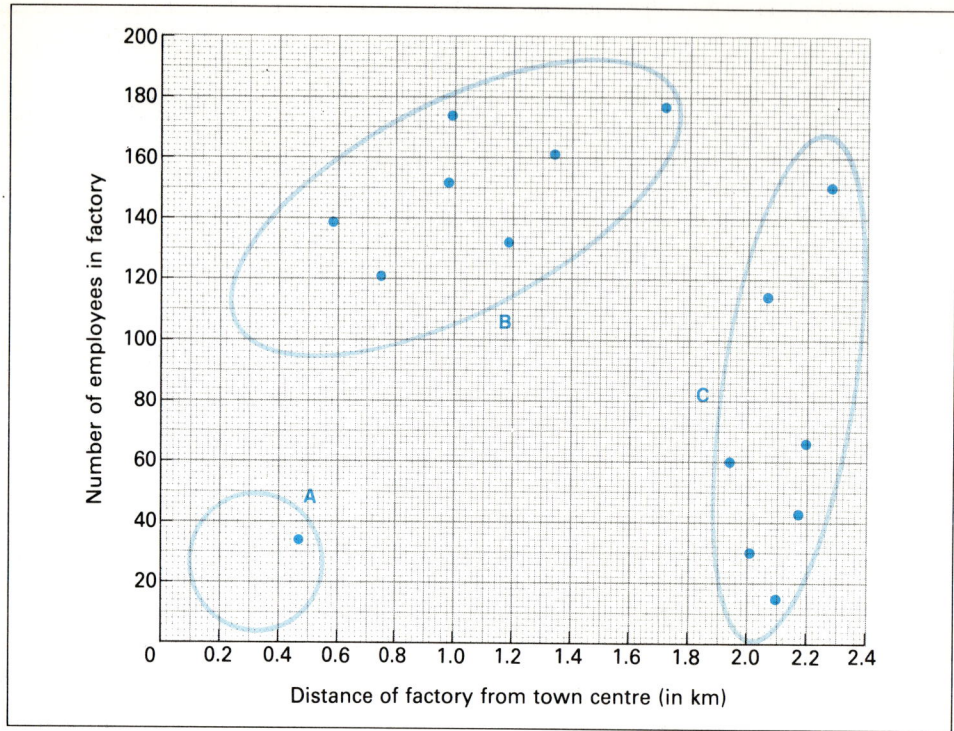

Figure 2.25

Number of employees in factory

Distance of factory from town centre (in km)

4 a Copy the scatter graph above, taking *great care* to plot the dots correctly.

b Plot a further ten dots on your graph to show this extra information:

Distance of factory from town centre (in km)	Number of employees in factory
0.2	17
0.2	30
0.3	12
0.3	40
0.3	116
0.4	101
0.4	97
0.4	134

c Which of the dot groups (A–C) on your graph are most likely to be:

☐ An old canal or river bank industrial area?
☐ A modern industrial estate in the suburbs?

Give reasons for each choice you have made.

5 a On a base-map of a town in your local area, plot the locations of the following types of industrial activity; you will need to use a different symbol/colour for each type. This information can be obtained at first hand and/or using 'Yellow Pages' or local business directories.

building, building
  materials
chemicals
electronics, electrical
  goods
engineering (except ships
  and vehicles)
food and drink

metals
paper, printing;
  publishing
ships, aircraft
timber, furniture
vehicles
others (if necessary)

b Also use colour to highlight any major features (e.g. rivers, railways, main roads and by-passes) which are likely to have influenced the location of these industries.

c Add a key to show the meanings of all the symbols and colours you have used; your map must also have a linear scale, for distances.

d Name the important features such as the town centre, rivers, canals and industrial estates.

e Carefully describe the *distribution* of industrial units shown by your completed map. (Hints: consider distance from town centre, any tendency for factories to cluster [group together], the location of the main industrial areas relative to water and transport features.) Add *explanations* for any interesting patterns you have noticed.

f In which ways are the industrial location patterns in your chosen town *similar* and *different* to those in Lancaster?

2.5 Redevelopment and renovation

You should now know that large inner areas of our industrial towns date from the nineteenth century; also that many of the older houses in these areas are

now well past their useful lives. Some of them are in a poor state of repair, while others are too cramped or lack basic facilities such as an inside bathroom and toilet (Fig. 2.26). Obsolete (out-of-date) housing has been one of the major problems of urban Britain in the present century.

In the years after the Second World War, British town planners believed the best way to tackle the problem of **urban renewal** was large-scale demolition, followed by total re-building. It is estimated that about 75 000 dwellings were pulled down each year during the 1950s alone. This type of drastic action was called **comprehensive redevelopment**. It gave the planners a golden opportunity to provide more recreational open space, improve shopping facilities and up-date road networks. Major highways could be built to ease traffic flow into the Central Business District, and inner ring-roads added to allow through-traffic to by-pass the central areas completely (Fig. 2.27).

▲ **Figure 2.26** Inner city housing in Liverpool

▶ **Figure 2.27** Comprehensive redevelopment in central Glasgow. Many of the older residents believe the planners have 'taken the heart out of Glasgow'

Comprehensive redevelopment proved to be very costly. It also wiped out some housing which was quite satisfactory. More important still was the way in which it shattered whole communities whose identities and friendships had taken many years to develop. The original inhabitants were either rehoused on large new estates on the **outskirts** of the towns, or encouraged to move to **new towns** even further away. This disruption has led to serious social problems in recent years.

In the early 1970s, planners throughout the country began to re-think their policies on urban renewal. They realised that high-rise blocks of flats which were erected shortly after the last war were beginning to create more problems than they had solved. Their occupants complained of feeling 'cut-off', and old people couldn't cope with the isolation. Parents were unable to supervise their children playing outside. Regular callers such as postmen and milkmen hated making deliveries within the tall blocks, especially when the lifts broke down! It also became common knowledge that suicide rates in high-rise developments were generally higher than for communities living at lower levels. Some of these blocks are now being demolished (Fig. 2.29) – less than 30 years after they were completed.

The planners' new policy – called **renovation** – used demolition more sparingly. Only the worst pockets of housing would be pulled down; any dwellings which were structurally sound were to be inspected so that they could be modernised and perhaps enlarged. This policy has been applied to both council and privately owned accommodation. Local authority 'home improvement grants' were paid to owner-occupiers wishing to bring their houses up to modern standards. This approach worked out far cheaper. It had the added benefit of keeping well-established communities intact, and preserved much of the character of our inner cities (Fig. 2.29). The next unit, based on the Toxteth district of central Liverpool, provides examples of both approaches to urban renewal.

A third interesting trend has been for wealthy people ('yuppies') to buy large old properties in central areas then spend large sums of money improving them. This process is called **gentrification**, and many examples of it can be seen in the fashionable inner London Borough of Islington (Fig. 2.30). As you might expect, gentrification often makes the less well-off even more aware of deficiencies in their quality of life.

▼ **Figure 2.28** 'Going down!' Demolition of a 21-storey block of flats in Hackney Wick, London

▲ **Figure 2.29** 'Gentrified' housing in Toxteth

◄ **Figure 2.30** 'Gentrified' housing in Islington

1 Explain what is meant by:
comprehensive redevelopment
yuppy
renovation
gentrification
urban renewal

2 List the main advantages and disadvantages of:
a comprehensive redevelopment
b renovation
c gentrification

3 Write a poem about life in a high-rise block of flats. Your poem *could* begin –
'We live in number one-two-four,
One of the flats on the very top floor'.

4 a Study one local example of each of the three types of urban renewal (comprehensive redevelopment, renovation and gentrification). This could include questionaire-based work with residents to obtain their opinions of the housing they live in.
b Use a variety of techniques to record *and assess* your findings (e.g. written work, maps, labelled sketches, graphs based on questionnaire responses).

2.6 Urban renewal in Toxteth

Toxteth is one of Liverpool's inner residential zones. It lies between the city's Central Business District and the more recent and much more attractive areas of housing around Princes Park and Sefton Park to the south-east (Fig. 2.31). Toxteth is a zone of great contrasts – partly for historical reasons, but also because of the changes which have transformed it during the present century. In 1981, it was the scene of some of Britain's worst street rioting in recent years (Fig. 2.32).

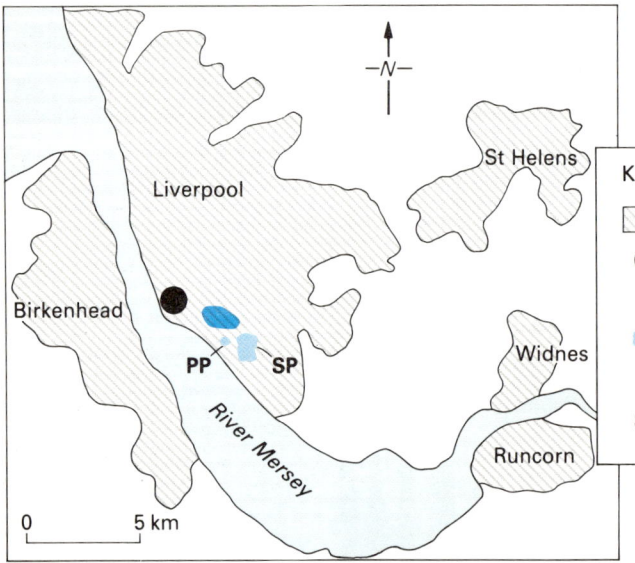

Key

◨ Built-up area

● Central Business District

◆ Toxteth

▢ Park

PP Princes Park

SP Sefton Park

◄ **Figure 2.31** The position of Toxteth

▲ **Figure 2.33** Princes Road, Toxteth

The Toxteth area was first developed during the nineteenth century, when Liverpool was becoming a major port and expanding very rapidly. The wealth obtained from its world-wide trade created a class of wealthy merchants; they could afford to build imposing detached villas and terraces of large town houses. These were sited along Princes Road (Fig. 2.33) a broad tree-lined avenue which runs through the heart of Toxteth down towards the parks. The port's activities also created thousands of extra jobs for working-class people. They were housed in parallel terraces of much smaller dwellings – on the side streets leading off Princes Road. In its heyday Toxteth bustled with activity and must have been a source of great pride to the city. The Toxteth of today is a very different place, for it shows all the scars of large-scale urban renewal. Although much of the area looks as though it suffered greatly in 1981, few of its houses were seriously damaged during the riots. Its present state is due much more to Liverpool's decline as a major port and the crippling unemployment which this has produced in the city.

◄ **Figure 2.32** Police extinguish burning tyres during the Toxteth riots in the summer of 1981

Some of the large houses built by the merchants have already been demolished. Many of those still used as dwellings have been sub-divided into flats and are presently occupied by people belonging to different **ethnic groups** (Fig. 2.34). A few have been renovated most successfully, and provide very attractive accommodation. Tragically, many more are in various states of decay; they have been empty and neglected for far too long (Fig. 2.35). Slates are missing from their roofs, their once impressive façades (fronts) are pitted, and gaping doors and windows allow wind and rain to rot their interiors. Unless drastic action is taken *very* soon, they too will have to be demolished.

Figure 2.36 shows one of the areas of smaller terraced housing which have now been comprehensively redeveloped. These dwellings had been hastily built of low-grade materials and so most of them have already been demolished. A few 7-storey blocks of flats were built about 30 years ago to re-house some of their evicted families, but these are already showing clear signs of wear and tear. The photograph in Fig. 2.36 illustrates the type of dwelling which is *now* being built on most of the cleared sites. Figures 2.37 and 2.38 show how one area at the northern end of Princes Road has changed over the last thirty years.

▼ Figure 2.37

▲ Figure 2.34   ▼ Figure 2.35

Answer the following questions, which are based on the text and *all* the illustrations in this unit.

1 Describe the position of Toxteth within Liverpool's built-up area.

2 a Say why both of these types of dwelling were built in Toxteth during the nineteenth century:

☐ detached villas and terraces of large houses
☐ much smaller terraced houses.

b Why has *each* type of dwelling deteriorated so much?
c What evidence of deterioration can be seen in Fig. 2.35?
d State which type of urban renewal (renovation *or* comprehensive redevelopment) has taken place on each of these streets:

☐ Princes Road
☐ Selborne Street.

3 Describe the recent type of housing shown in Fig. 2.36.

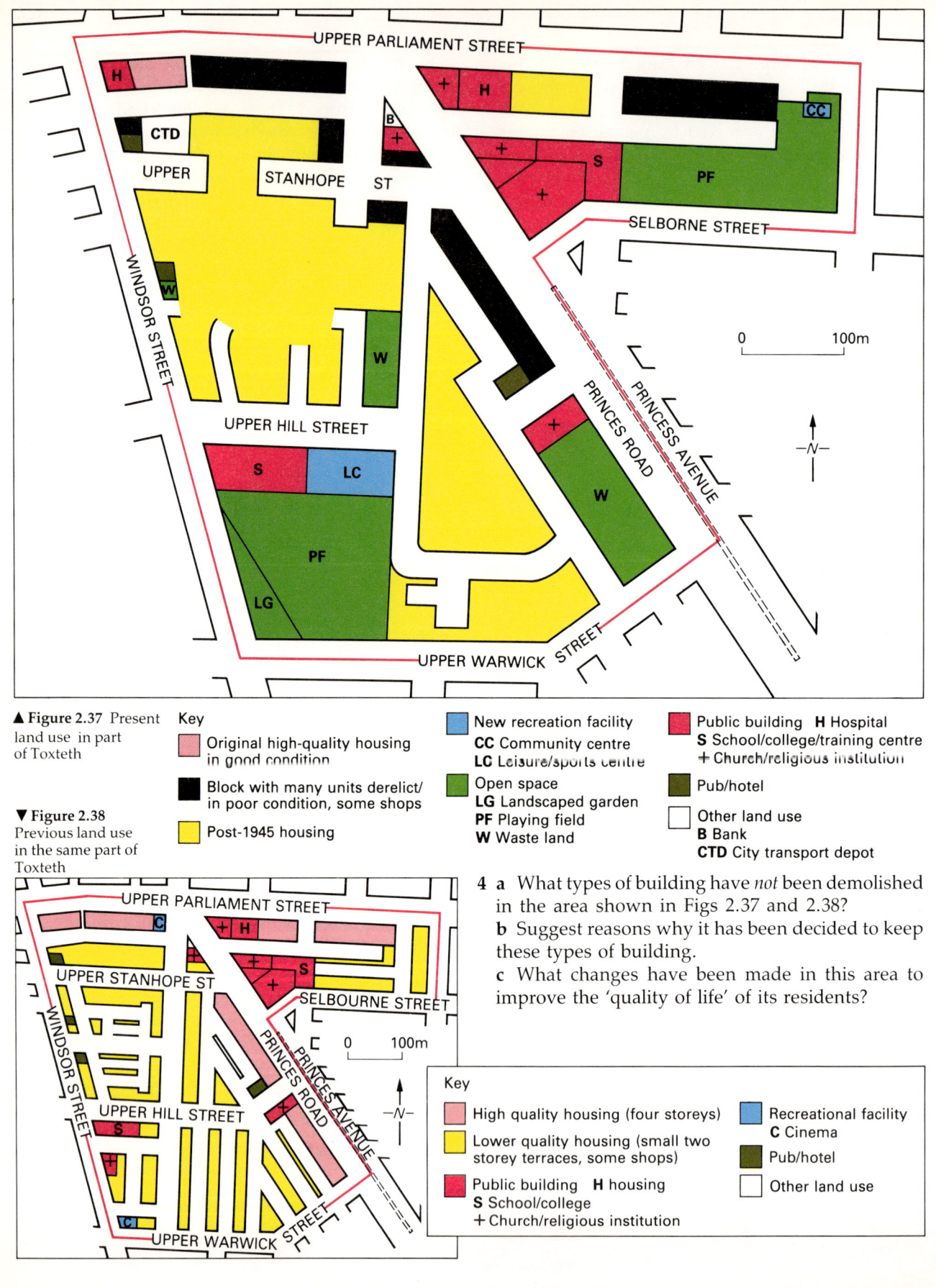

▲ **Figure 2.37** Present land use in part of Toxteth

▼ **Figure 2.38** Previous land use in the same part of Toxteth

Key

Original high-quality housing in good condition

Block with many units derelict/ in poor condition, some shops

Post-1945 housing

New recreation facility
CC Community centre
LC Leisure/sports centre

Open space
LG Landscaped garden
PF Playing field
W Waste land

Public building **H** Hospital
S School/college/training centre
+ Church/religious institution

Pub/hotel

Other land use
B Bank
CTD City transport depot

Key

High quality housing (four storeys)

Lower quality housing (small two storey terraces, some shops)

Public building **H** housing
S School/college
+ Church/religious institution

Recreational facility
C Cinema

Pub/hotel

Other land use

4 a What types of building have *not* been demolished in the area shown in Figs 2.37 and 2.38?
b Suggest reasons why it has been decided to keep these types of building.
c What changes have been made in this area to improve the 'quality of life' of its residents?

2.7 Local study of residential areas

It is best to study a town which is large enough to have some different styles of housing. Your teacher will then help you to choose the most interesting enumeration districts (or wards), and obtain the latest detailed census information for them. Your study may follow the pattern used in Unit 2.3 for Leyland – its final layout will of course depend on the kinds of information you decide to include, and how you wish to present the data. There are, for example, many ways of presenting the age–sex statistics tabled in Fig. 2.39. The population pyramid in Fig. 2.40 combines the numbers of teenagers in the '15' and '16–19'

year age groups (to keep to the five-year pattern used for every group except the last one), but you don't have to plot in such detail. You could double the size of each age group (giving 0–9, 10–19, etc.) or even double them again as in Fig. 2.16 on page 34. This simple formula shows how to convert numbers of people into the percentages needed for plotting *any* pyramid:

$$\text{Percentage of people in age/sex group} = \frac{\text{Number of people in group}}{\text{Total number of people in table}} \times 100$$

Age Range	Total persons in age range	Males		Females	
		SWD	Married	SWD	Married
0–4	67	41	–	26	–
5–9	44	20	–	24	–
10–14	34	21	–	13	–
15	9	4	–	5	–
16–19	18	11	–	7	–
20–24	67	16	13	14	24
25–29	83	13	30	7	33
30–34	50	5	24	–	21
35–39	29	4	11	4	10
40–44	19	2	7	3	7
45–49	17	3	9	1	4
50–54	16	–	7	1	8
55–59	23	2	12	1	8
60–64	16	3	4	–	9
65–69	9	–	6	2	1
70–74	7	1	1	2	3
75–79	6	–	1	4	1
80–84	1	1	–	–	–
85 +	–	–	–	–	–
Total	515	147	125	114	129

Note: 'SWD' stands for single, widowed and divorced.

▲ Figure 2.39

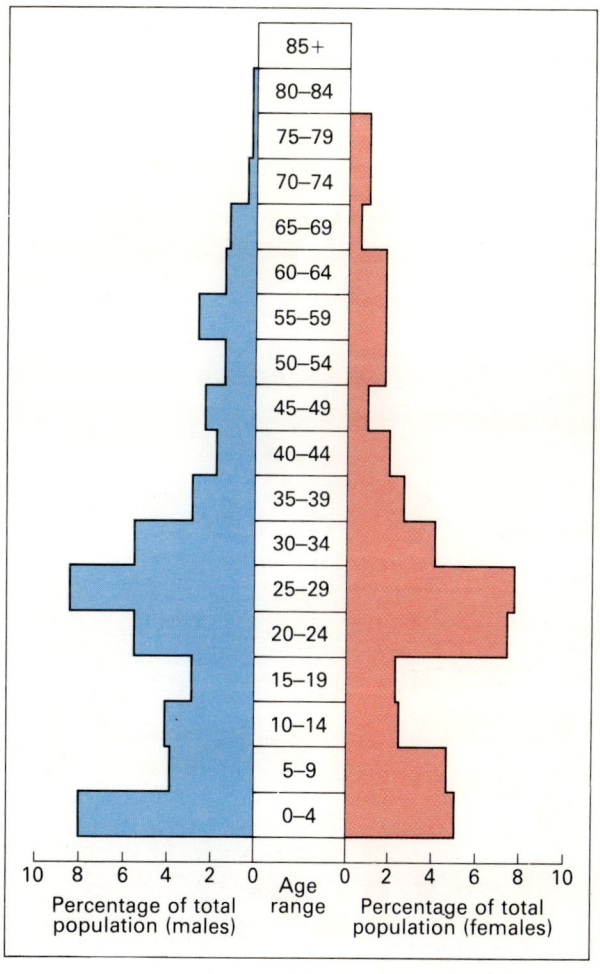

▲ **Figure 2.40** Population pyramid based on the information given in Fig. 2.39

It is often quite interesting to assess the physical condition of the housing in each district, and to compare it with the statistics provided by the census. These assessments should be based on *groups* of houses, since individual buildings may have been maintained to a different standard from those around them. The table in Fig. 2.41 lists seven types of information on which your assessment can be based, as well as the range of scores which may be awarded for each one (e.g. a group of houses showing few signs of 'paint peeling' could be given a score of 2).

Factor	Score for very serious condition	Score for perfect condition
Buildings tilted due to subsidence	0	9
Paint peeling	0	3
Roof sagging	0	7
Roof tiles missing	0	5
Wall surfaces chipped or crumbling	0	5
Windows broken or missing	0	7
Woodwork rotting	0	4

▲ **Figure 2.41**

The total score (out of 40) for each survey may be interpreted as in Fig. 2.42.

Range of scores	Description of general condition
35–40	Excellent
25–34	Generally satisfactory
15–24	Generally unsatisfactory
0–14	Drastic remedial action or demolition justified

▲ **Figure 2.42**

1 a Research the history of your chosen town, then briefly summarise what you have found out. You could include a line graph to show how its total population changed between 1801 and 1981 (the period for which British census figures are available).

b Draw a map similar to Fig. 2.15 on page 33 to locate the districts you have studied.

c Use tables and different types of graphs to display the census information you wish to include.

d Describe what each table and graph can tell you about the chosen districts, then, after class discussion, try to suggest reasons for what you have noticed.

e Carry out a 'physical condition' housing survey within each district (see Figs 2.41 and 2.42).

f Draw a sketch of each type of house you surveyed. Label your sketches to highlight all the defects revealed by the survey (see the example below).

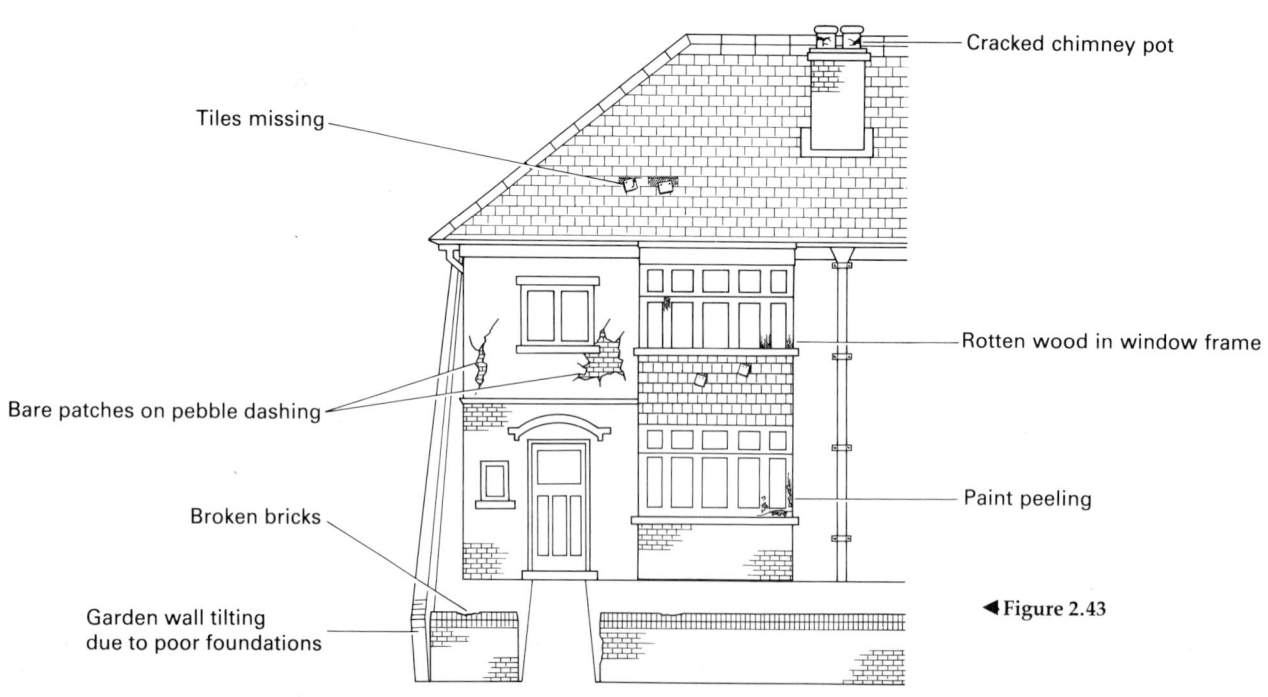

◀ **Figure 2.43**

2.8 Housing the people

Housing is one of the most interesting topics in geography for it can tell us so much about our way of life, both past and present. It fulfils one of our basic needs – that of shelter – and so providing it has been one of the main concerns of governments and individual families alike. Every country has its own way of meeting this need. Communist countries, for example, believe that housing is the responsibility of the state. Capitalist countries such as Britain encourage the private ownership of dwellings, but accept that not everyone can – or may wish to – buy their own house. Housing therefore tends to be a key issue for governments, and most political parties have very clear views on the subject. This is certainly true of the British Conservative and Labour parties (Fig. 2.44).

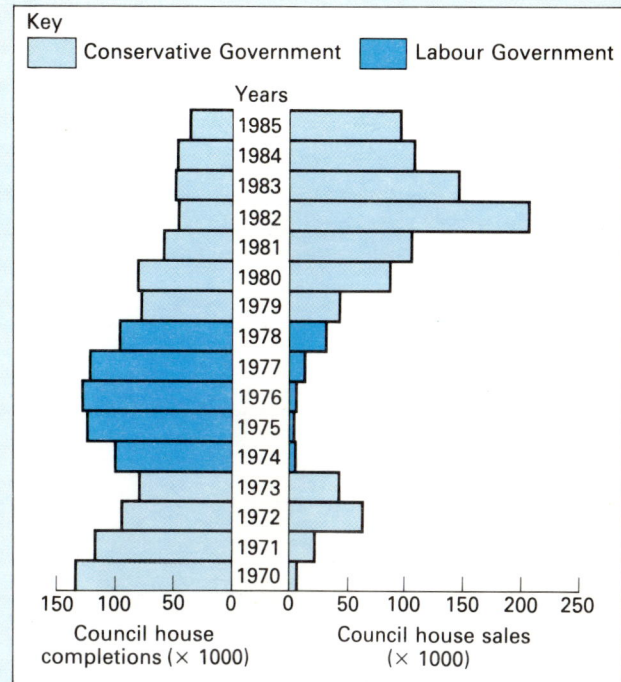

▲ **Figure 2.44** Horizontal bar graph showing the completions and sales of council houses in Britain, 1970–1982

◀ **Figure 2.45** New house prices in the United Kingdom relative to average male earnings

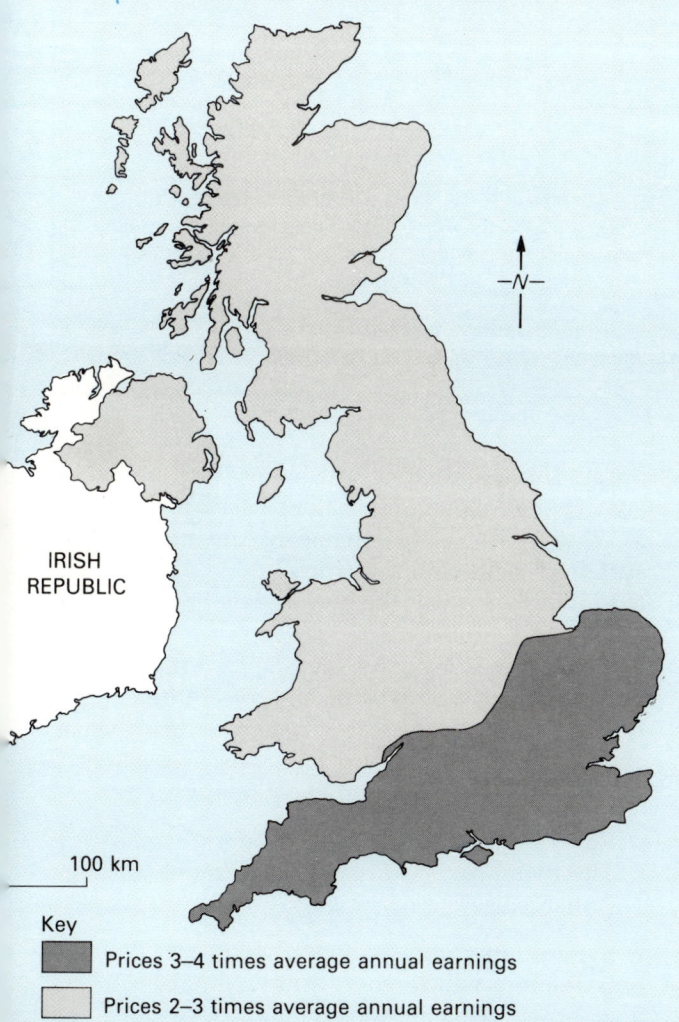

At present about 55% of all British householders are owner-occupiers; 32% rent property from local authorities; and 13% live in accommodation rented from private landlords. The following list describes some of the reasons why the proportion of *owner-occupied* dwellings is so high, and why it continues to increase in spite of widespread unemployment.

□ Buyers are able to repay the cost of their house by taking out a mortgage. The building societies were established to provide this service, but banks and local councils are also active in this way.
□ Mortgage repayments carry generous tax relief.
□ Improvement grants have been available since 1949 to subsidise the cost of up-dating old housing which is privately owned.
□ Since 1960, the government has supported the growth of 'housing associations'. These are groups of occupiers who have joint-ownership of all their properties.
□ Since the early 1980s, council tenants have been able to buy the property they are living in – often at a very attractive price.

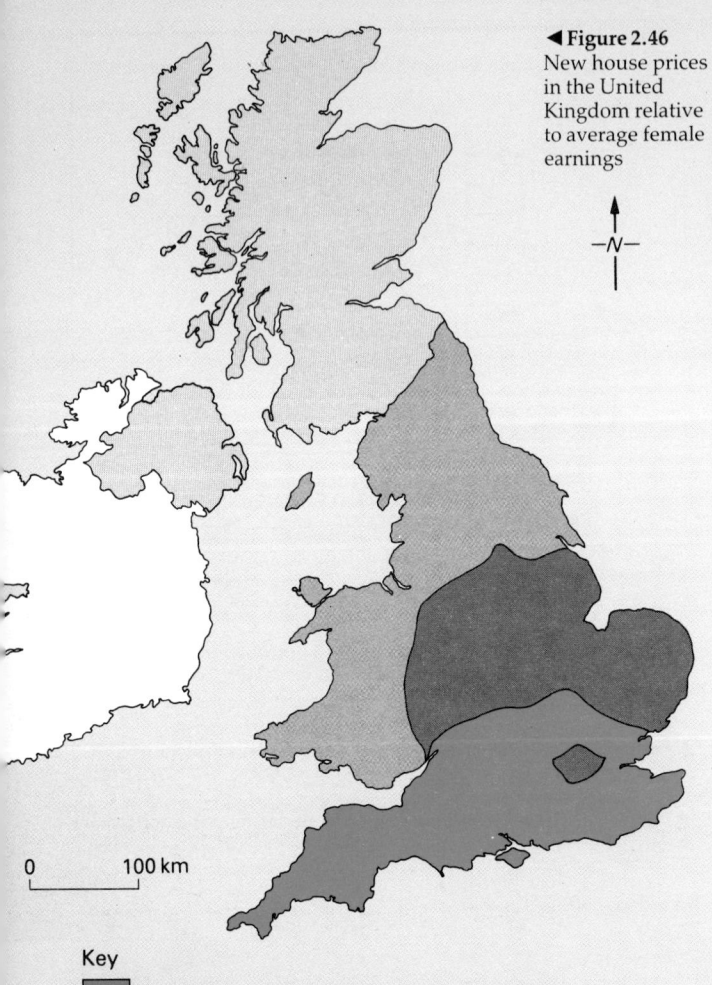

◀ **Figure 2.46**
New house prices in the United Kingdom relative to average female earnings

—N—

0   100 km

Key

▨ Prices over 5 times average annual earnings
▨ Prices 4–5 times average annual earnings
▨ Prices 3–4 times average annual earnings
▨ Prices 2–3 times average annual earnings

£50 instant homes may solve housing crisis

By Paul Stokes, West Country Correspondent

A BRITISH invention which could provide emergency accommodation for many of the world's 100 million homeless was unveiled by its creators yesterday.

The cheap and effective building system, developed by architects in Plymouth, Devon, involves panels which slot together and when bolted create a rigid structure capable of withstanding extreme weather conditions.

A simple one-person shelter can be erected within minutes at a cost of around £50. A four-person bungalow with kitchen, bathroom, lounge and bedrooms can be built in four days for £6,000.

Relief agencies have expressed interest in the shelter system, known as 'ownhome', for use after natural disasters such as earthquakes or hurricanes.

Another application being considered for the system is to create housing estates on inner-city wastelands at a fraction of the cost of conventional housing.

At the centre of the design is material using a sheet of steam-compressed insulation cork sandwiched between sheets of plywood.

'On their own they are of little value, but when held together by a special glue we have devised they have remarkable qualities for strength, durability and insulation,' said Mr David Terry, 48, the architect behind the idea.

Tests suggest the 'temporary' homes would last around 80 years, meet British energy insulation standards and be water and fire resistant to British safety standards when treated.

'We have spent two years creating, testing and using the board until we have reached the stage where we believe it could go a long way towards solving the world's housing crisis,' added Mr Terry.

▲ **Figure 2.47** Shelter systems for the homeless

1 What are the basic differences in the housing policies of:
a capitalist and communist governments?
b the British Conservative and Labour parties?

2 Explain why:
a the proportion of owner-occupied housing in Britain is both high and increasing,
b some people in the United Kingdom find it much more difficult than others to buy their own property. (Hint: Study Figs 2.46 and 2.47 in this unit, Fig. 4.5 on page 65, and these facts about 1986:

☐ The average rate of inflation was 3.5%.
☐ House prices in the London area rose by 26%.
☐ House prices in Northern England, Scotland and Northern Ireland rose by less than 3%.)

3 Study the newspaper article in Fig. 2.47 then answer the following questions based on it.
a What is the official name of the new shelter system described in the article?
b In which year was the new system first shown to the general public?
c Suggest reasons why the shelter system is so much cheaper than traditional types of housing.
d In what way can the new system be described as very 'flexible'?
e Why is the new system likely to appeal to:

☐ Relief organisations such as Oxfam?
☐ The planning and finance departments of large British cities?

f Suggest any other groups of people or types of organisation which could show interest in the new system. Give reasons for any suggestions you make.

3
Urbanisation Abroad

3.1 Introduction: urbanisation and Third World cities

It has already been stated that the populations of many Third World countries are increasing very rapidly. Because they are in poor countries, these cities cannot afford to meet even the basic needs of all their people. This unit outlines some of the many problems which result from their inability to cope. It ends with the personal story of one of South America's millions of migrant city-dwellers – the people who are largely responsible for the present scale of these problems.

A person's most basic needs are food, water, clothing and shelter. Ideally, these needs should be met by the efforts of individual families. In most *developed* countries, wages are usually high enough for employed people to feed and clothe their families quite adequately. Many can also afford to buy their own homes. All house-holders are expected to pay rates to their local councils who then use this money to provide services such as sewage disposal, a safe and reliable supply of drinking water, and public transport. Certain groups of people may find it very difficult to meet the high cost of living in these countries, hence the need for state pensions and unemployment benefits (Fig. 3.1). These are financed by taxing company profits and the people in higher income groups. Not all the inhabitants of developed countries enjoy a satisfactory standard of living, however, as Fig. 3.6 shows all too clearly.

▲ **Figure 3.2** General view of a typical CBD – in this case São Paulo, Brazil. Tall blocks of offices can be found in most large cities in the world.

The economic situation in a typical Third World country is generally very different. Both trade and industry are less well developed and so there are fewer companies able to create wealth and pay high wages. To make matters worse, the number of people who have to share a country's income (its GNP) is increasing all the time, and often very rapidly. Those without a reasonably paid job and who are in real need of financial help far outnumber the people who are employed and could support them through paying taxes.

We must be careful not to have too one-sided a view of life in Third World cities. Wealthy people *do* live in them, and some are extremely rich. This is obviously true for there are stores selling luxury goods such as perfumes and beautiful clothes, expensive cars jostle amongst the traffic and some districts have immaculate, splendidly-furnished flats and houses. Would you say that the city in Fig. 3.2 is in a developed or developing country? (It is not always easy to tell!)

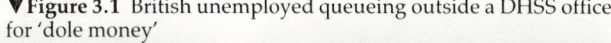
▼ **Figure 3.1** British unemployed queueing outside a DHSS office for 'dole money'

The fact that most of the richest people live in the cities is a very powerful attraction for the struggling country-dweller. The newcomers hope that their luck will change for the better once they reach the city, and that they will be able to get a secure, well-paid job there. Such jobs belong to what is called the **formal sector** of employment and they account for well under half of all employment in the Third World. A high proportion of the rural to urban migrants can't cope with the better-paid jobs in this sector. They do not have the necessary education or industrial skills for work in banks, offices and factories; many of them are **illiterate** (see glossary: **literacy**). These people have to seek other types of work within this sector, such as cleaning, or labouring on building sites.

The unemployed have therefore to create their own opportunities to earn a living – in the **informal sector**. Cleaning shoes, trading on the streets, washing cars and cutting hair are examples of this category of work (Fig. 3.3). In Calcutta, some of the more educated people who cannot find suitable work have offered their services as letter-writers and form-fillers to the illiterate. Lastly, there are those outside both of these sectors of the economy, who may be driven to stealing or prostitution to feed their families.

The effects of large-scale immigration on the cities can be very serious and visitors from developed countries are faced by sights which they will remember for the rest of their lives. The housing stock, for example, is totally inadequate, and many newcomers have to live in grossly overcrowded, decaying buildings near the city centre (Fig. 3.4). The only alternative for these people is to build their own homes on plots of vacant land. These are often little more than shacks made out of cardboard, straw, pieces of scrap wood and even discarded oil drums pressed into flat strips.

Such 'homes' offer little protection against wind and rain and their inhabitants are likely to develop ailments such as tuberculosis as a result. They give little privacy so that tempers fray very easily; divorce rates are generally much higher than in the rural areas. Social unrest is inevitable and bitterness and frustration sometimes lead to riots and revolutions against the government. The most destitute people of all live – and die – on the streets with only scanty clothes for protection against the weather (Fig. 3.5).

▲ **Figure 3.3** Shoe-cleaning on a street in Bihar, India, one example of the 'informal sector' of employment in Third World cities

▼ **Figure 3.4** Over-crowded housing in Singapore

► **Figure 3.5** Living – and dying – on Calcutta's streets

▼ **Figure 3.6** Sleeping rough in London. The number of homeless – and particularly young – Britons seems to be on the increase, and reminds us that developed country cities also have serious problems

The shacks already described usually cluster together to form **shanty towns** on cheap land around the city's edge (Fig. 3.7). In some places waste tips, cemeteries and even harbour jetties exposed to the sea have been built on by the squatters. The least attractive shanty towns are where the people are in fear of eviction (being forced to move elsewhere). The quality of self-built housing often improves quite quickly once this fear is removed. The unstable shacks are then replaced by more solid brick structures, but this can only be done after the necessary money has been earned.

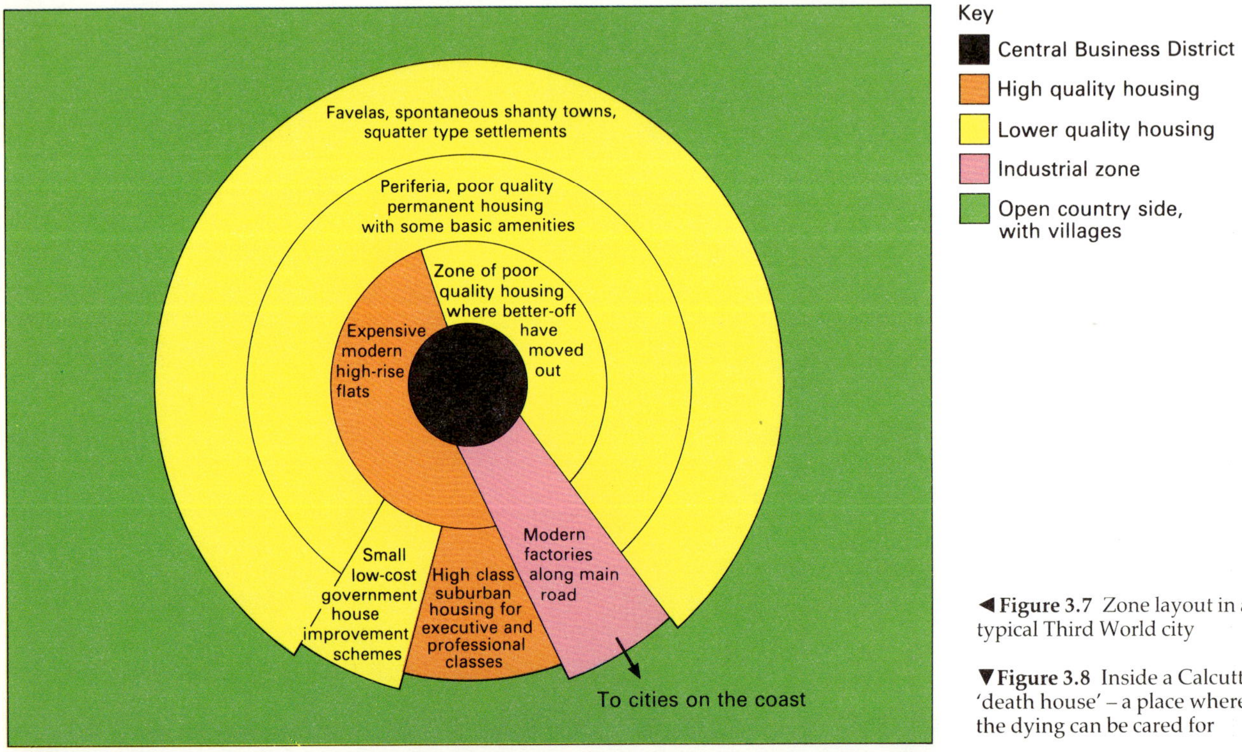

Key

- Central Business District
- High quality housing
- Lower quality housing
- Industrial zone
- Open country side, with villages

Favelas, spontaneous shanty towns, squatter type settlements

Periferia, poor quality permanent housing with some basic amenities

Zone of poor quality housing where better-off have moved out

Expensive modern high-rise flats

Small low-cost government house improvement schemes

High class suburban housing for executive and professional classes

Modern factories along main road

To cities on the coast

◄ **Figure 3.7** Zone layout in a typical Third World city

▼ **Figure 3.8** Inside a Calcutta 'death house' – a place where the dying can be cared for

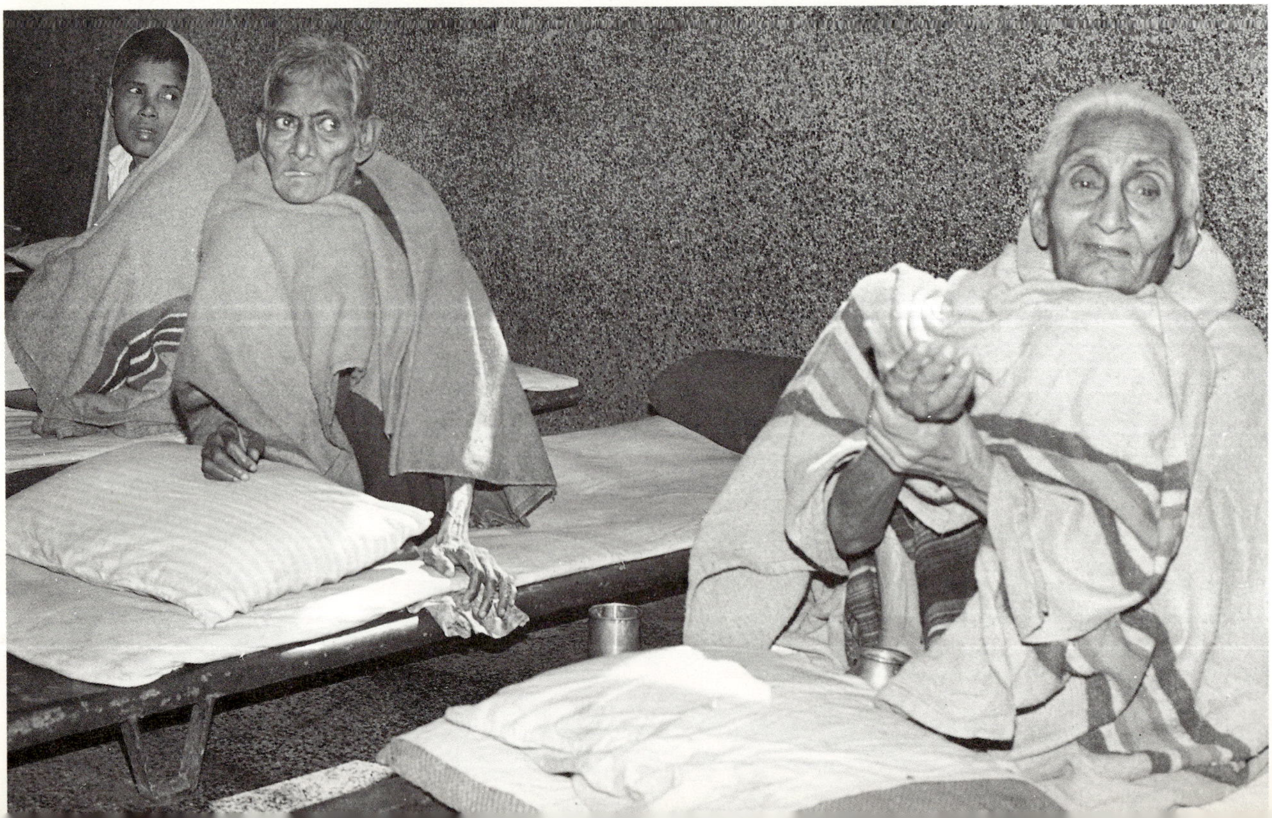

Many of the services which we take for granted are also totally inadequate in many Third World cities. The lack of proper sewage disposal facilities means that waste fouls the most overcrowded areas – ideal breeding conditions for diseases such as cholera (Fig. 3.8). Equally worrying is the absence of safe drinking water in many districts, and it is often left to the inhabitants to make their own arrangements to get supplies. Public transport is often grossly overloaded at peak travelling times and there isn't enough money to meet the demand for schools. It is quite common for parents to have to pay for all the books and pencils their children need at school.

This story describes what happened to the family of a Brazilian boy called Pedro when they moved recently from their village in the countryside to the great city of Rio de Janeiro. Parts of the story are quite depressing, but it does show what is happening every day to thousands of families throughout the Third World.

'Moving to the city was difficult, and nearly all our savings went on bus fares. We did try to get a free ride once but the driver spotted us and made us get off. He was really angry about it. Luckily, he had to leave us to sort out an argument on the bus, which was very overcrowded.

'We arrived at the edge of Rio late one night. We had been walking for a long time and were very tired. It had been raining and the tracks were muddy. The only lights we could see were on the main roads and in the tall blocks of flats near the city centre. We soon found out that is where the rich people live. None of my friends has ever been inside one because there are security guards on all the entrances. I don't think they trust people like us!

'Someone told us where my uncle lived but it took us the rest of the night to find his hut. We didn't have a place of our own and hoped he could put us up for a while. Uncle lived in a shack made of pieces of scrap wood nailed together. The roof leaked and he had put a bucket on the floor to catch the drips from it. We felt very sorry for him. He had lived there for three years and had only just got a job cleaning some offices near the docks. It didn't pay very much, but at least it kept him from starving. We chatted for a long time because we had lots of family news to catch up on. Uncle had been too ashamed to tell my dad how difficult things had

been for him since he came to Rio, and he can't write very well anyway. We told him about the droughts and how they had ruined our crops year after year. He went very quiet when mum said how scared she had been of the moneylenders and how we had to sell our land to pay them back.

'The next day, we all walked around the favella, which is the name we Brazilians have for our shanty towns. The rain had stopped but the little channels which ran between the houses were still quite full. The water looked very dirty and all kinds of rubbish floated in it.

'Not all the houses were as bad as uncle's. Quite a few were made of concrete blocks and had corrugated iron roofs. They belonged to the lucky ones who had managed to get jobs at the new car factory and could afford to have electricity supplied to their houses through overhead cables. Most people didn't have electricity, and they didn't have proper sewers and a supply of clean water either. Someone had saved up enough to pay for a water pipe to be laid to the favella. He made a living by selling water by the bucketfull. The City Council can't afford to give every house its own water supply, but they wouldn't do that anyway because they know the people have no right to be living on this land. I even heard one story about some huts near here being bulldozed by council workers. The people were furious about it, but couldn't stop it happening.

'The next day, uncle showed dad the best places to look for work. They went to the docks, the market place and the railway goods depot. Dad didn't get a job because they had already been taken by people who had started queuing very early that morning. He visited all three places every day for two weeks, but never got any work.

'Dad has now decided we have to move nearer the city centre so that he can join the job queues much earlier and won't have to pay bus fares to get to them. He has found a small patch of waste ground near the main railway line, where we hope to build our own hut. I'm dreading living there, but what else can we do?

'I'm going to start clearning shoes outside a posh new hotel right in the centre. It will be a bit boring and probably won't pay very much, but it might help me to meet some rich people. If I'm very lucky, one of them might take a liking to me and offer me a proper job! I'll just have to wait and see . . .'

1 **a** Write a summary of Pedro's story so far, in not more than 350 words.
b Now write a similar number of words to add a suitable ending to his story. It is up to you whether the ending is a happy or a sad one.

2 Make a simple 'line-drawing' of the scene below, then add detailed labels to it which show:

☐ the different types of material used to build the houses.
☐ the appearance of the houses *and* the area generally.

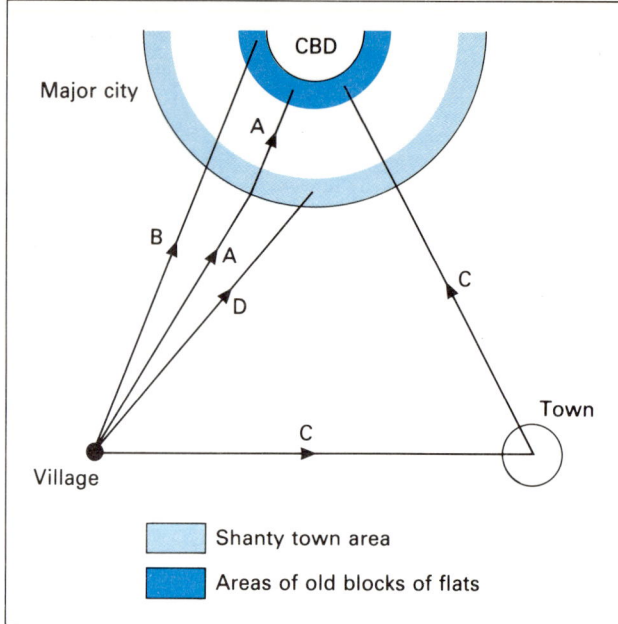

▲ **Figure 3.10** Common rural to urban migration routes

▲ **Figure 3.9** Shanty housing area in the industrial coast town of Puerto Supe, Peru

3 **a** Copy Fig. 3.10 which shows four journeys taken by many migrant families in the Third World.
b Which route is most likely to be taken by:

☐ A family which arrives at the city already having had some experience of urban life; this means that they are not unduly worried by overcrowding and heavy traffic?
☐ A family which has just sold its land; this means that the farmer and his wife need to find new jobs? They have saved up enough money to pay for a cheap flat.
☐ Pedro's family (according to the story you have read)?

4 For each statement below:
a say whether it is *true* or *false*,
b give as much information as you can to justify your answer in a.

Statement 1 Shanty towns are areas on the outskirts of Third World cities where all inhabitants live in very poor housing conditions.

Statement 2 Most of the problems of Third World countries would disappear quite quickly if all their adults had secure, well-paid jobs.

Statement 3 The districts in and near to the centres of Third World cities are very similar in some ways to those in developed country cities, but are quite different in others.

3.2 Urban renewal in the Third World

A shortage of decent housing is one of the most serious problems facing the world today. The developed countries can go some way towards solving their housing problems as they have much greater resources with which to build new accommodation and improve the old. However most large cities in both developed and developing countries do have pockets of ageing, sub-standard accommodation and these are usually located just outside the Central Business District (Fig. 3.11).

Developing countries have a higher proportion of inadequate housing. This is because they are poorer, but also because their populations are increasing much faster. This unit studies just two ways in which Third World countries are trying to improve housing standards. You will see that Caracas has had a wider choice of options, made possible by the income created by Venezuela's exports of crude oil. Zambia too receives much of her income from the export of a raw material (copper), but this trade has not had the same impact on its capital – Lusaka.

▶ **Figure 3.11** 'Skid Row' – a ghetto area in one of New York's inner residential zones

▼ **Figure 3.12** High-rise development in Caracas, Venezuela

Housing developments in Caracas

The population of Caracas has trebled since 1950. It is estimated that a quarter of its 2.5 million people still live in self-built houses grouped together in shanty towns which the Venezuelans call 'ranchos'. These occupy at least one-fifth of the city's total area. While few of them have adequate sewage facilities, most do have electricity supplies and safe drinking water.

Caracas' housing policy in recent years has been to build tall blocks of flats (Fig. 3.12). Venezuela was once a Spanish colony and it has copied the high-rise type of development used in Europe since the Second World War. Equally important is the site of Caracas itself. The city nestles in a valley surrounded by steep mountains, and so flat building land is very scarce. The profits from oil have helped the city to build modern shopping centres, office blocks and road networks; all these have taken up much of the best building land, forcing the people to live in blocks like the ones in the photograph, Fig. 3.12.

Housing developments in Lusaka

Caracas and Lusaka are similar in a number of ways. Lusaka is the capital of Zambia, a country which was once part of a great European empire (the British). It too is growing very rapidly due chiefly to rural to urban migration. Two important *differences* are that Lusaka has plenty of flat land on which to expand, but far less money to invest in new housing.

Shortly after gaining independence from Britain in 1964, Zambia started a number of building projects to re-house some of its poorer people. It soon realised that the European methods were so costly that only a few people could be re-housed by using them. The Zambian government therefore had to think of a much cheaper alternative. It hit on the idea of what became known as the **site and service schemes**. These were basically self-help programmes, in which the people would do most of the building work them-selves, but the government would give them some money, some expert advice and some security on the land they occupied.

The alternative scheme proved to be a great success, and has been copied by many other countries. It did not happen by accident; the Zambian government had noticed that the people who lived in the shanty towns around Lusaka had steadily improved their properties. Many of these were quite well construct-ed of sun-dried bricks, and had been plastered and painted to make them more pleasant to live in. The *problem* was that these areas lacked many basic facilities and were becoming seriously overcrowded.

One of the greatest worries was the practice of digging water wells very near to the pits used for burying household waste – an obvious health risk. Another was the state of the 'roads' within the shanty towns. These were merely dirt tracks which quickly became unusable during the rainy season. The government has tackled all these problems, and even offered sums of money to compensate families who volunteered to leave overcrowded areas. This would give the remaining families extra room to enlarge their houses as more children were born. Figure 3.13 gives you a clearer idea of how a typical shanty town area on the edge of Lusaka has changed with the help of the government.

1 a Trace an outline map of South America. On it shade the area of Venezuela, label this country with its name, then locate and name the position of Caracas.
   b Produce a similar map of Africa to show the loca-tions of Zambia and Lusaka.

2 Describe the many factors which have influenced the recent housing policies in both Caracas and Lusaka.

3 a Using your own words as far as possible, describe the types of housing development which have taken place recently in Caracas and Lusaka.
   b State which of these housing developments you consider to be best suited to most Third World countries.
   c Give reasons for your answer to b.

4 a Copy out this table, then complete its fourth column after dividing the Gross National Product figures in the second column by the population figures shown in the third.

Country	Gross National Product in Millions of US$	Population	Per capita GNP in US$	Annual rate of population increase
Developed countries:				
Australia	148 064	15 226 000		1.3%
United Kingdom	523 256	56 459 000		0.1%
Developing countries:				
Venezuela	60 028	14 714 000		3.0%
Zambia	3 837	5 680 000		3.0%

b Discuss the information given by your com-pleted table, then write a summary of the diffe-rences between the developed and developing countries included in it.

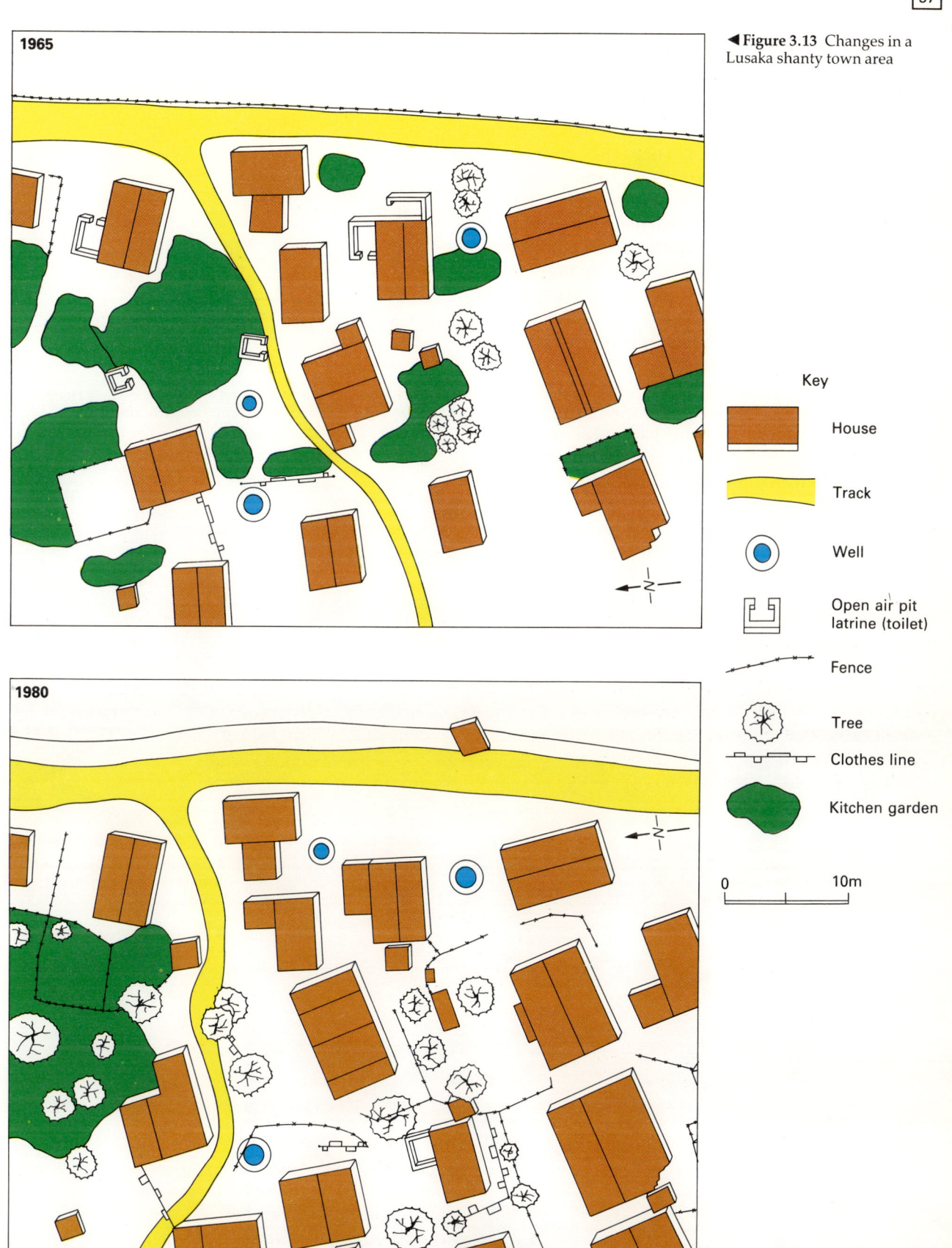

◄**Figure 3.13** Changes in a
Lusaka shanty town area

1965

1980

Key

House

Track

Well

Open air pit
latrine (toilet)

Fence

Tree

Clothes line

Kitchen garden

0   10m

3.3 Housing developments in Hong Kong

Many of the world's developing countries are too poor to provide most of their people with adequate housing. The British colony of Hong Kong has been more fortunate and this unit describes the measures it has taken over the last thirty years to improve its housing stock. The colony is a flourishing port and industrial centre. Low wages have helped it to manufacture goods cheaply and sell them abroad at very competitive prices. Electrical goods, clothing, toys and books are just a few of its many exports.

Hong Kong is one of the world's most overcrowded places. Its six million people live at an average **population density** of 5 000 per km^2 – *twenty* times higher than the figure for the United Kingdom. Its population grew very rapidly after the Second World War (Fig. 3.14). This was partly due to a high birth rate, but was mainly because of the immigration of homeless refugees from mainland Asia. Many of these newcomers came from China. More recently, others have fled from Vietnam as a result of the lengthy war which devastated that country. Most of them entered Hong Kong illegally (without permission) and they have put extra pressure on the colony's re-housing programmes.

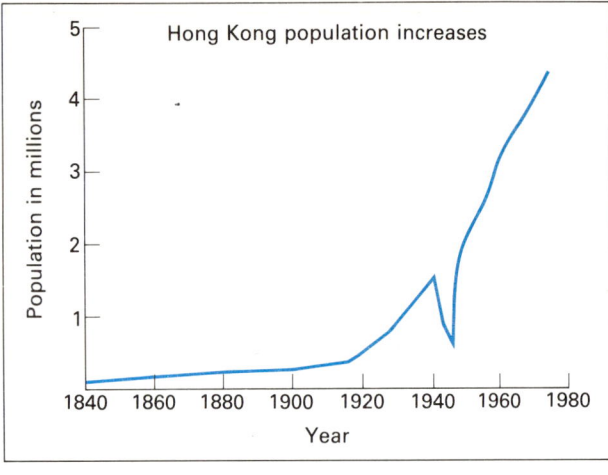

▲ **Figure 3.14** Line graph showing the dramatic rise of Hong Kong's population. Why did it drop suddenly after 1940?

▲ **Figure 3.15** Sampans form a floating town in a sheltered Hong Kong bay

◄ **Figure 3.16** Shanty dwellings on a Hong Kong hillside

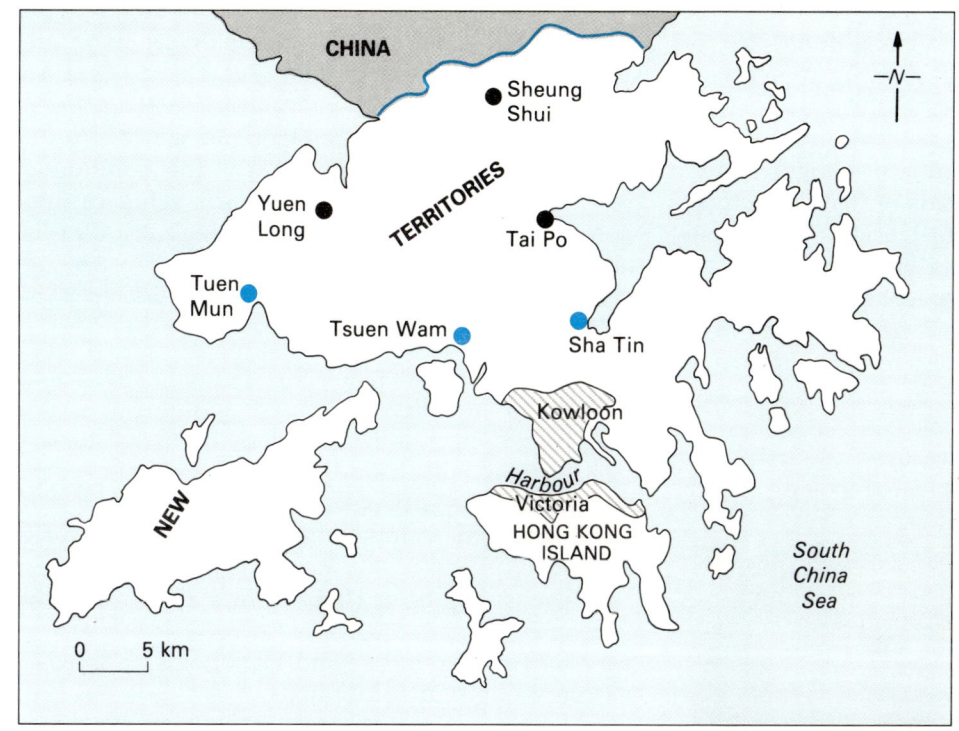

Key

◻ Long established built-up area

● New town

● Expanded town

∿ International boundary

▨ Chinese territory

◀ **Figure 3.17** Urban development in the British colony of Hong Kong. The 'New Territories' become part of China in 1997, when their lease expires

▼ **Figure 3.18** Part of Sha Tin – a new town in Hong Kong

Figures 3.15 and 3.16 show two types of accommodation still found in the poorer areas of Hong Kong. Flat land suitable for building on is so scarce that many people have had to build their shacks on steep hill-sides. Others live on sampans clustered together in sheltered bays.

The housing situation in Hong Kong was suddenly made much worse by a disastrous fire on Christmas Day, 1953, which left 50 000 people homeless. This tragedy shocked the government into doing more about the shortage of decent accommodation. It decided to build three new towns; these were sited on narrow strips of flat land by the coast, and Sha Tin was partly built on land reclaimed from the sea (Fig. 3.17).

The architects given the job of planning the new towns were asked to provide housing for about two million people, at an average density of 2 500 per km^2. At first, the architects achieved this density by building very small flats in blocks up to ten storeys high. More recent blocks have reached thirty storeys, but provide larger accommodation. There are wide open spaces (Fig. 3.18) between many of the blocks, where the people can meet and relax. Each new town has at least one racecourse – to cater for the Chinese love of gambling! As well as the new towns, Hong Kong has also enlarged a number of existing settlements.

People are re-housed in both types of town when they reach the top of the government's waiting list. The pressure on land is so great that many cannot be re-housed until the land they are living on has first been cleared! The evicted people are given temporary accommodation which consists of a concrete floor and an asbestos roof supported on wood uprights. There are no walls, but the families can use some of their small removal allowance to build temporary ones (Fig. 3.19). Fresh water and sewage facilities are provided free.

Industry has been the key to Hong Kong's success in the past, and so the architects were very keen to include some industrial estates in their plans. Tsuen Wan, for example, has many textile and clothing factories as well as a shipyard and a large chemical works. There are also hundreds of 'flatted factories' in the new towns. These are blocks – often many storeys high – like the one in Fig. 3.20. They have large areas of open floor space which individual companies can sub-divide and equip as they wish. Most of these smaller companies have fewer than ten workers and specialise in making plastic goods, toys, shoes, clothes and cheap jewellery.

In spite of all these measures, thousands of squatters still live in dilapidated-looking shacks clinging to the hillsides. In fact many of these are surprisingly well-equipped inside. This is because plenty of unskilled work is available in Hong Kong, and both children and grandparents are able to increase their families' incomes. It is this continuing need for workers which still attracts refugees into the colony. Hong Kong is paying a high price for its industrial success, and now faces the prospect of having to build more new towns just to keep pace with the housing problem.

1 Correct these statements by changing the words which are in italics.
a Hong Kong is a British colony on the east coast of *Australia*.
b Hong Kong is peopled mainly by *Britons*.
c Most of the colony's flat land is on *Hong Kong Island*.
d *Tourism* has been Hong Kong's chief source of income.
e Sha Tin, Tsuen Wan and Yuen Long are three of Hong Kong's *expanded* towns.
f The Hong Kong government built the new and expanded towns because they *had a great deal of money to spare*.
g Flatted factories are *large single-storey buildings*.
h Large open spaces have been provided between the new blocks of flats for *use as car parks*.
i Some of the squatters' shacks are well-equipped inside *due to the generosity of the government*.
j Hong Kong has now *completed its house-building programmes*.

2 a Describe the type of temporary housing provided by the Hong Kong government.
b Suggest ways in which this accommodation might be improved.

3 Complete these statements by adding as much information as you can.
a Hong Kong's population has increased very rapidly because ...
b The colony is so densely populated because ...
c The Hong Kong government had little option but to spend large sums of money on new housing because ...
d Temporary accommodation for re-housed families is usually necessary because ...
e The new housing developments have included some industrial units because ...

▲ **Figure 3.19** Temporary accommodation in Hong Kong. Concrete floors and wooden uprights support an asbestos roof.

▼ **Figure 3.20** Flatted factories in Hong Kong

3.4 Moscow: communist capital city

Politics often influence major planning decisions, especially those affecting primate cities. One of the best examples of this is Moscow – the largest and (since 1918) the capital city of the USSR. The Soviet Union is a communist developed country, which means that it belongs to the Second World (see page 7). It also means that its politicians can dominate planning matters far more effectively than those in capitalist countries. For example, *all* Soviet factories are owned by the state and their workers can expect to be told where they shall live. Moscow shares many of the problems facing other modern cities, and so it is quite interesting to see how communist thinking has influenced the way in which they have been tackled.

Year	Area of Moscow (in km²)	Population of Moscow (in millions)	Average population growth rate (per year)
1811	50	0.27	
			0.8%
1871	90	0.60	
			4.1%
1912	177	1.62	
			1.8%
1926	228	2.03	
			9.5%
1939	294	4.54	
			1.7%
1959	356	6.04	
			1.5%
1970	886	7.06	
			1.5%
1981	886	8.20	

▲ Figure 3.21

Moscow's most formidable task has been to house a rapidly increasing population (Fig. 3.21), due chiefly to migration from rural areas. The results of this migration have been too few houses and serious overcrowding; a major cause was Stalin's Five Year Plans in the 1920s and 1930s which created large numbers of new jobs *in the cities*. The Soviet leader did not forsee the increasing food shortages in the major cities – especially in the depth of winter. The German invasion of the western part of the Soviet Union in 1941 reduced the immigration rate for a while but it quickly resumed after the war and by the late 1950s drastic action had to be taken to increase and improve the housing stock. This mainly took the form of 5-storey blocks of flats (Fig. 3.23), and the necessary extra land was given to the Mossoviet (Moscow City Council). Later blocks were much higher and used modern pre-fabrication methods. Under this system, the various sections are mass-produced in factories then taken to the building sites for assembly.

▲ **Figure 3.22** Joseph Stalin, whose name meant 'man of steel'. This ambitious and ruthless leader laid the foundations for the mighty Soviet Union of today. The efficient Moscow underground railway network was largely his idea

▼ **Figure 3.23** Much of the post-war housing in Soviet cities looks very similar. Five-storey blocks were usually chosen as they were considered low enough to be built without lifts. This makes large areas of the capital rather formal and uninteresting to look at

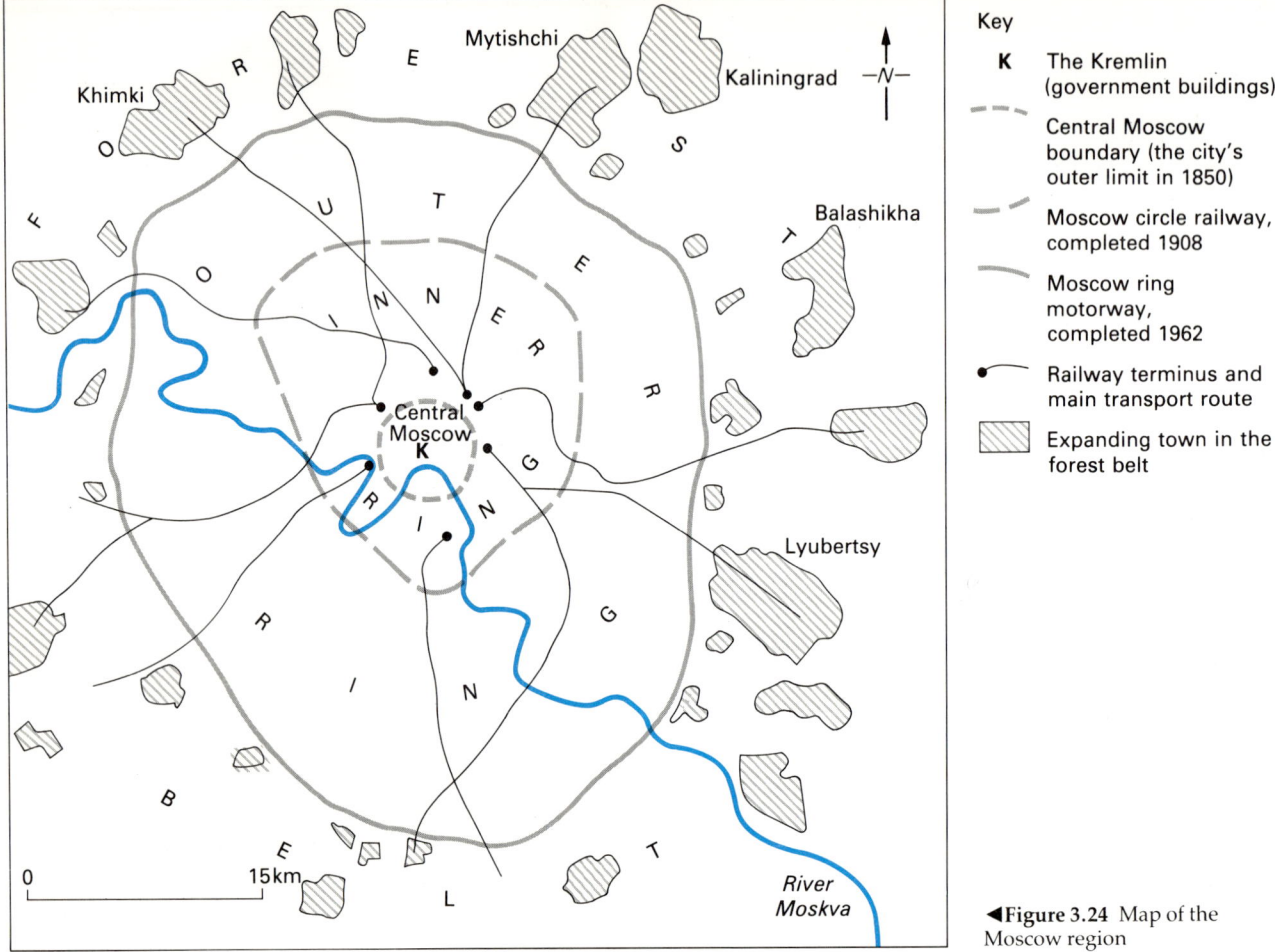

Key

K The Kremlin (government buildings)

- - - Central Moscow boundary (the city's outer limit in 1850)

– – Moscow circle railway, completed 1908

—— Moscow ring motorway, completed 1962

•— Railway terminus and main transport route

▨ Expanding town in the forest belt

◀**Figure 3.24** Map of the Moscow region

The urban sprawl which took place in the late 1950s/early 1960s formed Moscow's 'outer ring' (Fig. 3.24). This lies between the circle railway and the 109km long ring motorway. It is dominated by housing of the pre-fabricated type described on page 61, although more recent blocks have tended to be much taller. The Mossoviet has allocated its housing stock in such as way that 'richer' and 'poorer' districts do not develop easily. Housing equality has been a major communist aim ever since the 1917 revolution, and the public ownership of accommodation has made this possible in most urban areas. Cheap and efficient city transport has also helped to produce uniformity in residential districts. Figure 3.24 shows that the most recent urban development has taken place in a fourth and even larger area, which lies beyond the ring motorway. This used to be densely forested but is now becoming increasingly urban as its many 'new' towns also sprawl relentlessly outwards.

1 Describe Moscow's:
a site (see Fig. 3.24)
b situation (use an atlas to obtain Moscow's general position within the USSR).

2 **a** Draw a line graph to show how Moscow's population changed between 1811 and 1981.
b Between which years did Moscow's population increase at the highest *rate*?
c State the reasons for the very high rate of increase during this period.

3 Describe the communist influence on Moscow's development. Figure 3.23 should prove especially useful to you, but you must also read the text with great care.

International Migration

4.1 Introduction: migration into Britain

People have been migrating (moving) into Britain for thousands of years. Their numbers increased sharply during the last two centuries – due to great improvements in international transport which made it easier to travel long distances. Many Irish, for example, came during the nineteenth century to find work building canals and railways. Few black immigrants entered Britain during the 1800s, although some Chinese did establish laundries and restaurants here and quickly became known for their high standards of service (Fig. 4.1).

Immigration into Britain peaked after the Second World War (which had left our economy very depressed). The government decided to encourage immigration as a way of increasing our workforce and re-building our industries. It advertised in India, Pakistan, the West Indies, and many other British territories overseas. Financial help was offered to those who wished to come here.

This campaign was highly successful; it produced a *net* gain of nearly 400 000 people during 1960–62 alone. In fact it was so successful that the government passed the Commonwealth Immigration Act in 1962 to limit immigration *from these countries* in future years! By the late 1960s, the number of black immigrants had dropped to less than 30 000 per year, and many of these were relatives of immigrants already living here. Britain had become a truly **multi-racial** (mixed) **society**, and Race Relations Acts were passed in 1965 and 1968 to protect the ethnic minority groups in this country. These Acts made it an offence to discriminate (act) against people because of their colour, religion or way of life.

Britain's inward and outward migrations have been much more stable in the 1970s and 80s. We continue to accept limited numbers of migrants, and have tried to be sympathetic to refugees wishing to escape persecution in their own countries. Our entry into the Common Market in 1973 has encouraged the (often temporary) migration of workers between member countries, leading to a net *emigration* in some years. This pattern of short-distance migration seems certain to increase, especially after the completion of the Channel Tunnel.

1 a After class discussion, suggest *what* transport improvements have made immigration so much easier in the nineteenth and twentieth centuries.
b Name two groups of nineteenth-century migrants to Britain.
c Why did these migrants come here?

2 a Name three *Asian* countries from which immigrants came to Britain in the 1960s. (Hint: two are named in the text; un-jumble BASHDELANG to find the third.)
b Which of these islands are in the West Indies (between North and South America)?

Barbados   Madagascar   Trinidad
Jamaica   Seychelles
(Hint: use an atlas!)

3 Explain what is meant by the terms *multi-racial society* and *minority groups*.

4 Re-word these statements to make them read true:
a The British government encouraged immigration to help some of the countries in the Commonwealth.
b The peak period of black immigration into Britain was 1945–50.
c The Commonwealth Emigration Act of 1960 reduced black immigration into Britain.
d The Race Relations Acts have helped to reduce immigration to Britain.

◄ Figure 4.1

4.2 Immigrant zones in urban areas

This unit aims to answer the questions 'Where do immigrants into Britain live – and why?' The simplest answers are 'In the cities – for work', for recent census figures show that about two-thirds of them live in the major conurbations. Greater London alone has 40% and the West Midlands, around Birmingham, about 15%. At the other end of the scale, most settlements with fewer than 50 000 have very small immigrant communities.

Newly arrived immigrants have always tended to move into the largest settlements. During the nineteenth century, large numbers of Irish immigrants were attracted to Liverpool and Glasgow, as these ports were easy to reach. In the 1930s, European Jews adopted the Golder's Green district in north London as their main 'base' in southern England. Post-war immigrants from the Commonwealth have also been drawn to the cities, and west London's Southall was one of the first areas to be settled by this group (Fig. 4.2).

▲ **Figure 4.2** Southall – a popular immigrant area in West London

▲ **Figure 4.3** Immigrant bus drivers – two of many employed by London Transport. Some of them are second generation immigrants, which means that *they* were born in Britain

▶ **Figure 4.6** The Notting Hill Carnival, in North London

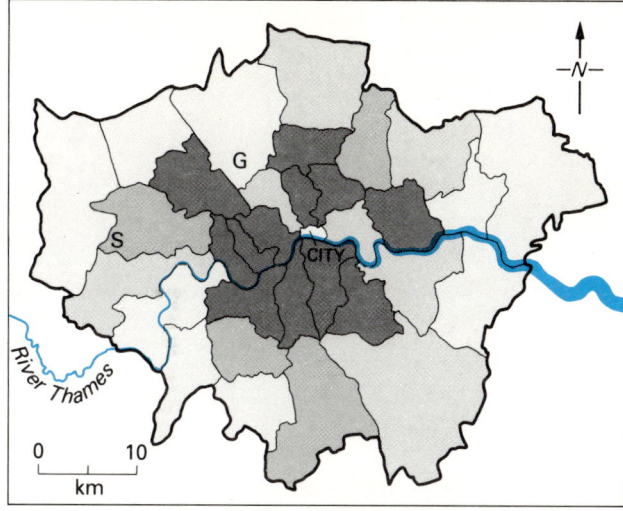

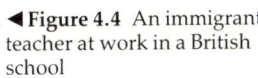

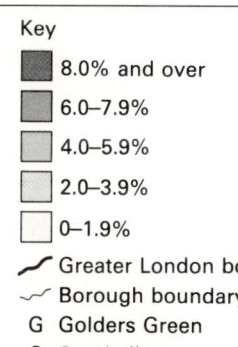

◀ **Figure 4.4** An immigrant teacher at work in a British school

▲ **Figure 4.5** Immigrant concentrations in London's boroughs

Key

■ 8.0% and over
■ 6.0–7.9%
□ 4.0–5.9%
□ 2.0–3.9%
□ 0–1.9%

ꜱ Greater London boundary
ꜱ Borough boundary
G Golders Green
S Southall

group. Friendship ties give them a feeling of security in their first years in this country; relatives may be able to offer advice, financial help and perhaps temporary accommodation. Immigrants often have a far more caring attitude towards others than we do. For example, it is usually their custom to look after older relatives, rather than place them in 'homes for the elderly'.

Newcomers can relax and settle down to life in a strange country much more quickly if they are amongst people who share the same language, religion and traditions (Fig. 4.6). Most ethnic communities arrange their own social activities, and these make it easier for recent arrivals to meet and later choose a marriage partner from their own ethnic group. Being amongst one's own people also gives some measure of protection against intimidation by rival sections of the community. This is especially important during periods of high unemployment, when money is scarce and tempers are easily roused.

1 Draw a bar graph to show the percentages of recent immigrants in Greater London, the West Midlands and 'other areas' of Britain (see page 64).

2 Why do immigrants tend to settle:
   a in cities and large towns?
   b in inner residential areas?
   c in close-knit communities?

3 **a** Name two districts in London where immigrants have clustered together, and the ethnic groups which have favoured these districts.
   b Where are these two districts located within the Greater London area, according to Fig. 4.5?

4 List the various reasons why immigrants tend to cluster together in certain urban areas. This can be done under two headings:
   a Reasons which *force* them into clusters.
   b Reasons which are largely a matter of *choice* (e.g. those concerned with their traditional way of life).

5 After class discussion, suggest any *disadvantages* of immigrants grouping together in close-knit communities (e.g. education, employment opportunities, links with other groups of people).

6 Make a display of cuttings taken from local and/or national newspapers which examine the quality of life of immigrant communities in Britain. The most interesting sections of these cuttings could be highlighted by underlining or fluorescent-colouring them. You might also add your *own* comments and reactions to what the cuttings have to say.

Cities hold many attractions for new arrivals. They are major centres of employment and so are most likely to provide job opportunities for them. Many immigrants were farmers in their own countries, but others brought valuable skills and experience with them (Fig. 4.4). Very important too is the availability of reasonably-priced accommodation. Our conurbations still have large areas of nineteenth-century housing, as well as tall blocks of flats – hence their popularity with newly-arrived immigrants (Fig. 4.5). Furthermore, immigrants often have great difficulty buying houses in the more modern and more attractive residential zones. It may not be easy for them to convince estate agents and building societies that they can repay large mortgages; also the white residents of these areas may put pressure on their neighbours not to sell to immigrants.

There are many social reasons why immigrants prefer to live close to other members of the same ethnic

4.3 Belfast: a divided city

Belfast – the primate city of Northern Ireland – has been called a 'divided city' in recent years. This is because many of its residential zones are occupied mainly by Catholics *or* Protestants. This is indirect contrast to the cities in Great Britain, in which people usually decide where to live for economic or social (rather than political or religious) reasons. This unit discusses some of the historical reasons which have led to the **segregation** (separation) of the religious communities in this city.

Early plans of Belfast show that some segregation was taking place as early as 1685. Belfast had already become the chief port for trade with England and Scotland, and this encouraged people from both of these *Protestant* countries to seek work there. Many of these immigrants worked on the docks and in mills which wove flax into fine linen (Fig. 4.7). Later, Belfast became an important ship-building and engineering centre and these industries attracted more Protestant immigrants. They also encouraged thousands of Irish farmers – who were *Catholics* – to leave their land to seek better-paid jobs in the city.

Unrest between the two religious communities grew steadily worse, and even the partition (division) of Ireland into two separate countries failed to ease the tension. If anything, it made matters worse. The new republic in the south of the island (called Eire) would not accept that the province of Ulster (in the north) should remain part of the United Kingdom.

The size of the migration into Belfast put great

Figure 4.7 Linen – once an important industry in Northern Ireland

pressure on the city's resources. Poverty, overcrowding and eventually widespread unemployment were the result. These led to much jealousy and bitterness between the Catholic and Protestant communities. Rival mobs clashed openly on the streets, property was looted and there were frequent attacks on individuals. The present 'troubles' began in 1969 and the murders and bombings frequently reported in the press are now organised by 'para-military groups' within the province (Fig. 4.8).

▲ **Figure 4.8** A scene from 'the troubles' in Northern Ireland

Economic planning regions	Percentage of Unemployed Adults
South-East England	8.2
East Anglia	9.1
South-West England	8.6
West Midlands	9.7
East Midlands	13.2
Yorkshire and Humberside	10.8
North-West England	12.9
Northern England	13.6
Wales	15.9
Scotland	13.4
Northern Ireland	17.8
UK average	11.7

▲ **Figure 4.9**

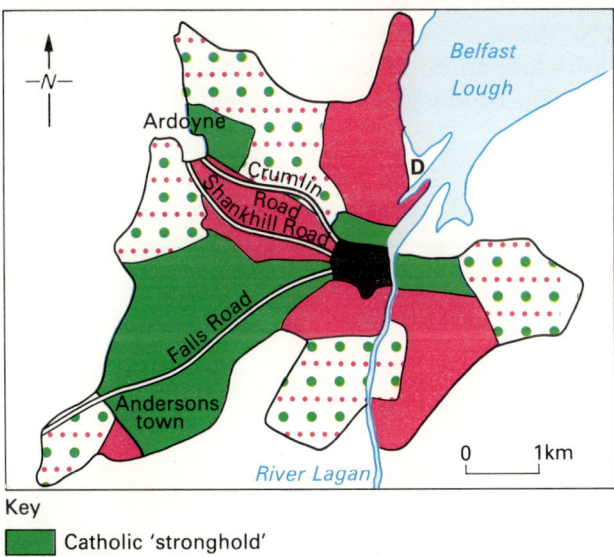

▲ **Figure 4.10** Harland and Wolff shipyard, Belfast

Key
- 🟩 Catholic 'stronghold'
- 🟥 Protestant 'stronghold'
- ⬜ 'Mixed' area
- ⬛ CBD
- **D** Docks
- ⌒ City boundary

▲ **Figure 4.11** Segregated areas in Belfast

Unemployment has certainly helped to increase the tension in recent years. Figure 4.9 shows that Northern Ireland has the highest rate of unemployment of all the UK's Economic Planning Regions. This is the result of the steady decline of the city's main industries. For example the workforce of shipbuilders Harland and Wolff (Fig. 4.10) contracted from over 25 000 in 1970 to about 5 000 in the mid-1980s. The linen textile industry has virtually collapsed due to competition from cheaper man-made fibres.

It is hardly surprising that both Catholics and Protestants huddle together in their 'own' residential areas for protection. Thousands of families have migrated *within* Belfast to produce the segregated distribution shown in Fig. 4.11. Internal migration on this scale has been possible because large areas of old terraced housing have been demolished and replaced by new estates on the outskirts of the city. One such suburb is Andersonstown – a Catholic 'stronghold'. The inner city areas also have similar patterns of segregation. Two streets which run through both inner and outer residential areas and are now widely known are Falls Road and Shankhill Road. The so called 'Peace Line' created in 1969 between these two

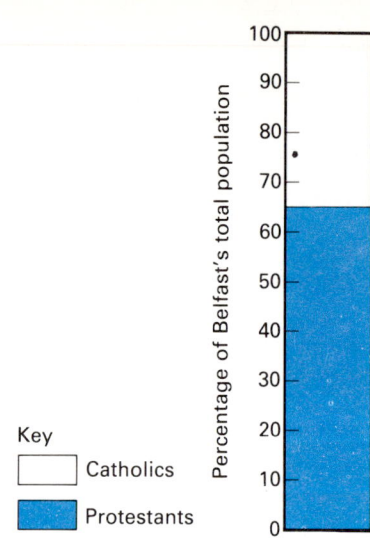

Key
☐ Catholics
■ Protestants

Percentage of Belfast's total population

◀ **Figure 4.12** The 'Peace Line' divides Catholic and Protestant areas in central Belfast, between Falls Road and Shankhill Road

▲ **Figure 4.13** Divided bar graph showing the division of Belfast's population into Catholics and Protestants, 1981

roads is shown in Fig. 4.12 which proves that its appearance is in fact anything but peaceful!

1 After class discussion, write sentences to explain the meanings of these words:

bitterness   partition
jealousy   province
Eire   republic
looting   segregation
para-military group Ulster

2 State *briefly* why Belfast has been described as a 'divided city'.

3 Which of the following Belfast streets and housing areas are:
a Catholic strongholds?
b Protestant strongholds?

Andersonstown   Falls Road
Crumlin Road   Shankhill Road

4 a Copy the divided bar graph above.
b Draw a similar graph for 1790, when 8% of Belfast's population were Catholics.
c What changes are shown by these two graphs?

5 a Illustrate the information given in Fig. 4.9 by shading an outline map of the United Kingdom's eleven Economic Planning Regions

You *could* use: yellow for under 10%,
            red   for 10.0%–14.9%,
            brown for 15.0% and over.

Add a box key to show the meanings of your colours.
b State how Northern Ireland's percentage unemployment rate compares with that of:

☐ South-East England
☐ Scotland
☐ the whole of the United Kingdom.

c Give some reasons for Northern Ireland's very high rate of unemployment.

6 a Say whether this statement is *true* or *false*: 'Past migration is one of the *chief* reasons for the present troubles in Northern Ireland'.
b Give reasons for your answer to a.

4.4 Chinatown in San Fransisco

As you now know, recent immigrants to Britain tend to live in the inner residential zones of our towns, where housing is cheaper. This unit shows that immigrants to *foreign* countries are also attracted to this type of area and that they share many of the problems faced by newcomers to Britain.

The discovery of gold in California in 1848 brought floods of immigrants from overseas to the west coast of America. Like thousands of other newcomers, the Chinese either worked on the goldfields or helped to lay railway tracks across the Rocky Mountains. When the gold began to run out and the railways neared completion, most of these immigrants moved into the cities.

San Fransisco now has one of the United States' largest Chinese communities. It is concentrated in a group of 24 'blocks' near to the city's Central Business District (Fig. 4.14). 35 000 Chinese live there, many of them crowded into old sub-standard housing only two or three storeys high – a great contrast to the much taller apartment buildings usually found in North American cities. The serious overcrowding in Chinatown has led to a high rate of tuberculosis amongst its residents. The crime rate there is also higher than in the 'white American' districts of the city.

Few of the Chinese speak English particularly well,

▲ **Figure 4.15** A busy street in China town, San Francisco

and this makes it difficult for them to compete for jobs. They are not afraid of hard work, however, and many Chinese have started their own businesses (Fig. 4.15). Chinatown's restaurants are famous for the high quality of their food and service and are extremely popular with all sections of San Fransisco's population. Some of the Chinese have become millionaires – through sheer determination and hard work. The most successful businessmen usually move out of Chinatown into more pleasant areas on the city's outskirts.

1 Explain why a large Chinese community developed in San Fransisco.

2 Copy out the following statements which are correct, but change any false ones to make them read true.

  □ Chinatown is located near to the docks and the Central Business District.
  □ Chinatown's crime, tuberculosis and unemployment rates are much lower than the average for the whole of San Fransisco.
  □ Most of the Chinese live in tall, overcrowded apartment blocks.
  □ The Chinese are hard workers, and many of them have created their own employment opportunities.
  □ The Chinese' inability to speak English well has reduced their chances of obtaining highly paid jobs in the Central Business District.
  □ The wealthiest Chinese still prefer to live in Chinatown.

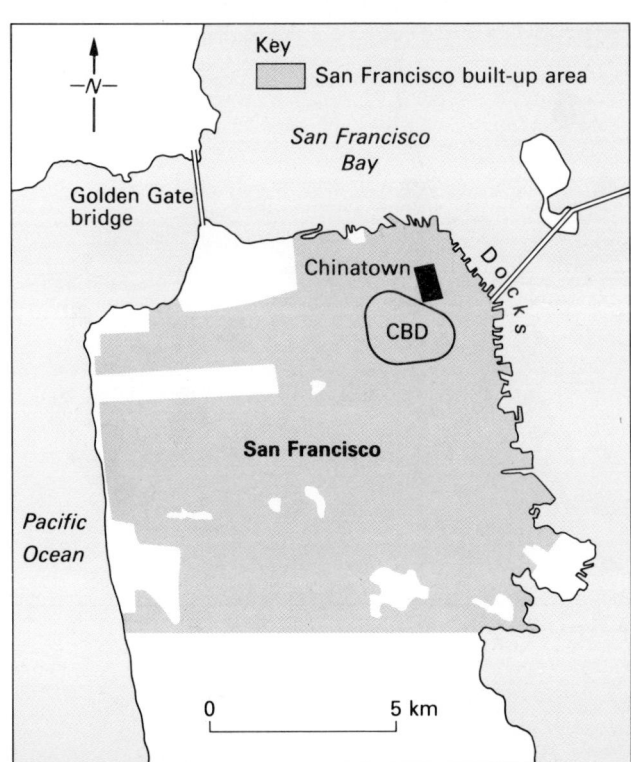

◀ Figure 4.14

4.5 Townships in South Africa

Counter-urbanisation usually takes place because people *wish* to move to a different residential area, not because the state has decreed where they shall live. The situation in South Africa, however, has been very different since the 1950s due to that country's policy of **apartheid** (pronounced 'apartight'). This has led to the almost total segregation of the various ethnic groups (Fig. 4.16) and produced many serious social problems in and around the major cities.

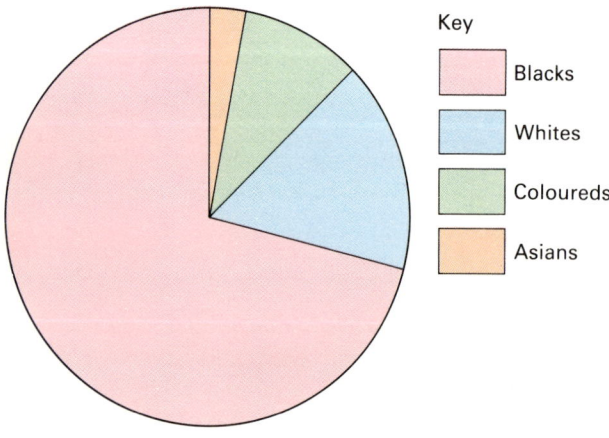

▲ **Figure 4.16** Pie graph showing the division of South Africa's population into its major ethnic groups

Key
- Blacks
- Whites
- Coloureds
- Asians

The white population first considered racial separation after great deposits of gold were discovered in the 1860s, close to where Johannesburg now stands. The whites controlled all the new mines there, but needed large numbers of black workers to extract the gold. This meant that the two groups *had* to live within travelling distance of the mines. Segregation in the Johannesburg area was eventually achieved by forcing the black population within the city to migrate outwards; these people had no alternative but to leave the more pleasant residential zones (which were to be exclusively for the whites) and move into special **townships** much further away from the city centre (Fig. 4.17). Soweto (the *South-Western Township*) is the largest of these and now has about two million inhabitants. The location of some townships means that blacks have to make very long journeys to work in the Central Business District; one-way journeys of more than two hours on crowded buses and trains are quite common – and costly.

Figure 4.18 shows one part of Soweto's urban sprawl. The standard type of house is quite small, often over-crowded, and lacks privacy. Each unit has its own toilet, usually located in a small hut at the bottom of the garden. Coal and oil are used for heating and lighting. Both of these fuels cause air pollution and on damp winter days the townships are often blanketed in **smog** – an unhealthy mixture of *smoke* and *fog*. The quality of life of the blacks in

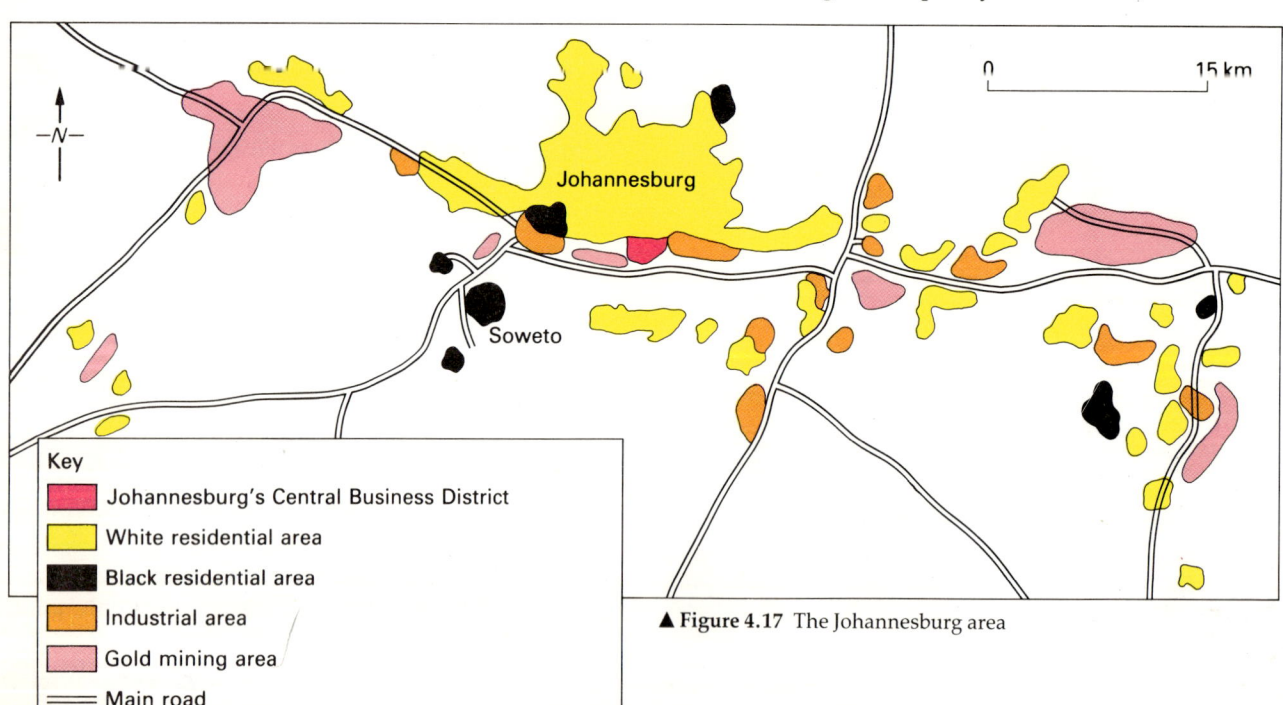

Key
- Johannesburg's Central Business District
- White residential area
- Black residential area
- Industrial area
- Gold mining area
- Main road

▲ **Figure 4.17** The Johannesburg area

these townships is much lower than that of the whites in areas like the one pictured in Fig. 4.19. The black population is deeply resentful about this and has been involved in much street violence during the 1980s. The South African government is only just beginning to improve the lot of non-whites in the country. For example, in 1986 it scrapped the hated 'pass laws' which had required blacks to carry a special pass at all times; the law even applied to famous athletes – while *actually running* in competitions! The measures which have been taken so far are 'too few and too late' as far as South Africa's non-white population – and world opinion generally – are concerned, and the social unrest within the country seems bound to continue for some years to come.

1 a Briefly, what does the word apartheid mean?
b Explain how the words Soweto and smog were first obtained.

2 Describe:
a The position of Soweto township from the CBD of Johannesburg (i.e. compass direction and distance, centre to centre).
b The general location of the black townships within the Johannesburg conurbation.
c The general location of the white residential areas within this conurbation.
d The general locations of *both* types of residential zone relative to the gold mining areas.

3 a Compare the overall sizes of the black and white residential zones within the conurbation.
b What is curious about your answer to a? (Hint: study Fig. 4.16 again.)

4 Contrast the two scenes in Figs 4.18 and 4.19, under two separate headings:
a The houses themselves.
b Their general environments, including garden areas.

▲Figure 4.18 Part of Soweto's monotonous sprawl

▶Figure 4.19 A 'white' suburb of Johannesburg; the smaller buildings house black servants

5

Shopping Facilities

5.1 Introduction: clustering of services

Clustering means the grouping together of similar land uses and activities within an urban area. It happens because clustering benefits the people who own *and* use the buildings in our towns and cities. For example, businesses have always tended to cluster in central areas. The CBD is not only conveniently located at the heart of a settlement; it also has direct transport links which radiate outwards to all the residential zones. In other words it has the most accessible location within a built-up area.

Localised clustering often takes place in smaller areas *within* the town centre. This is partly for the convenience of users, who wish to 'shop around' for the goods and services they need – without having to spend a good deal of time and money doing this.

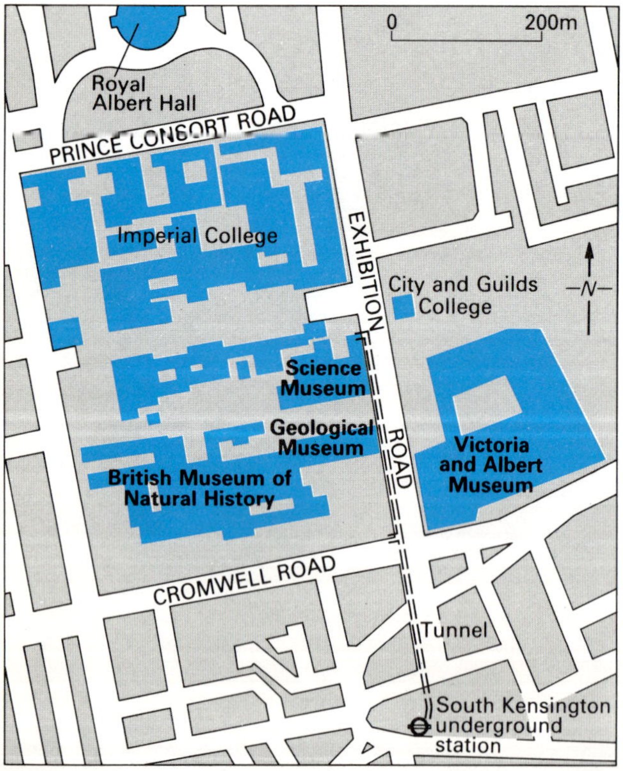

▲ **Figure 5.1** Museums 'clustered' around Exhibition Road in West London

Localised clustering has proved *essential* to certain types of business, those which cannot operate successfully without a constant and reliable supply of up-to-date information. Meeting regularly, face-to-face, develops mutual trust between business people in a way which telephone calls and letters alone cannot possibly do.

The example of localised clustering in Fig. 5.1 is taken from a western district in central London, but similar examples can be found in all large settlements. The localised clustering of banks, large stores and places of entertainment is a particularly common feature of the urban scene.

1 Explain carefully what the terms 'clustering' and 'localised clustering' mean.

2 Account for the localised clustering of the museums on and around Exhibition Road in central London.

3 Suggest possible advantages of localised clustering for each of these groups of workers:
 a bank managers
 b insurance agents
 c estate agents
 d jewellers
 e journalists

4 a Plot the distributions of post offices and insurance agencies on the street map of a town in your local area.
 b Which of these two distribution patterns appears to be *more clustered*?
 c Give reasons for the difference you have noted in b.

5 List a number of examples of *localised* clustering in your own/a nearby town. This could be done under two column headings:

Activities/services	Names of streets on which localised clustering takes place

5.2 Shopping facilities in Morecambe

This unit examines the various types of shopping facilities in Morecambe – a popular holiday resort on the Lancashire coast – and is based on the illustrations on the next two pages. Like most British towns, Morecambe has:

☐ a Central Business District
☐ suburban shopping centres
☐ parades of shops
☐ 'corner' shops
☐ recent shopping developments on the outskirts.

Morecambe developed from the ancient settlements of Bare, Heysham, Poulton and Torrisholme. Of these, Poulton grew the fastest because the main railway station was built there (in 1850). This made it the obvious choice for the CBD and now has branches of such chain-stores as Marks & Spencer, Boots, Tesco and Woolworth, an open-air market and a modern indoor Arndale Centre (Fig. 5.2).

Bare, Heysham and Torrisholme are now part of 'Morecambe's' built-up area, but their inhabitants still talk of 'going down to the village' when shopping there. They are all busy **suburban shopping centres**, as the list in Fig. 5.3 clearly shows.

The **parades of shops** are also to be found in the suburbs, but these have fewer businesses (usually 5–10). The parade shown in Fig. 5.4 is on the main road which links Heysham village to Morecambe CBD.

There are 'corner shops' throughout the town (Fig. 5.5), but those which depend on food sales are in decline – for the reasons stated in the next unit. They do however provide a convenient and friendly if somewhat expensive service for local residents.

The most recent shopping developments have taken place on the *outskirts* (Fig. 5.6). All are large units built on cheap, flat land. They are next to modern ring-roads and have plenty of space for customer parking. The largest development of all is an Asda **superstore** conveniently located on open land between Morecambe and nearby Lancaster (which have a combined population of 100 000). These new developments are proving very popular, and have taken much trade away from the Central Business Districts of both towns.

Morecambe's five shop locations can be rearranged into a **shopping hierarchy**. This means they are listed in order of size and importance, with the CBD and the corner shops at the top and bottom respectively.

It is also quite helpful to arrange the *goods* being sold in some sort of order. Expensive items such as dining room suites are called **high-order goods**. At the other end of the scale are the *low*-order goods (e.g. bread) which are much cheaper and bought very regularly.

1 After class discussion, rearrange:
a The five shop locations into a shopping hierarchy.
b These items into their correct order, starting with high-order goods:

cassette (or record) shampoo
coat washing machine
eggs

2 Which of the five shop locations are the following *most* likely to use:
a A family on a monthly shopping expedition; convenient, free car parking is desirable?
b A pensioner living in an old terraced house?
c A family wishing to furnish their new home, which is 10 km from the town centre?
d A mother living in a suburb, accompanied by young children; she doesn't have a car, but needs to visit a post office and buy some groceries and medicines?
e A DIY enthusiast, who wishes to buy a wide range of tools and decorating materials?

3 Write down at least five facts about each of the shop locations/groups in Morecambe.

4 a Carry out a survey of a suburban shopping centre in a town in your local area. Ideally, it should have about the same total number of businesses as the one in the Bare district of Morecambe.
b Use the information you have obtained to produce a table similar to the one in Fig. 5.3.
c After class discussion, compare the information given in both tables. (Hints: Which types occur in both centres? Which types are most common in both centres?)
d Suggest reasons for your comments in c above.

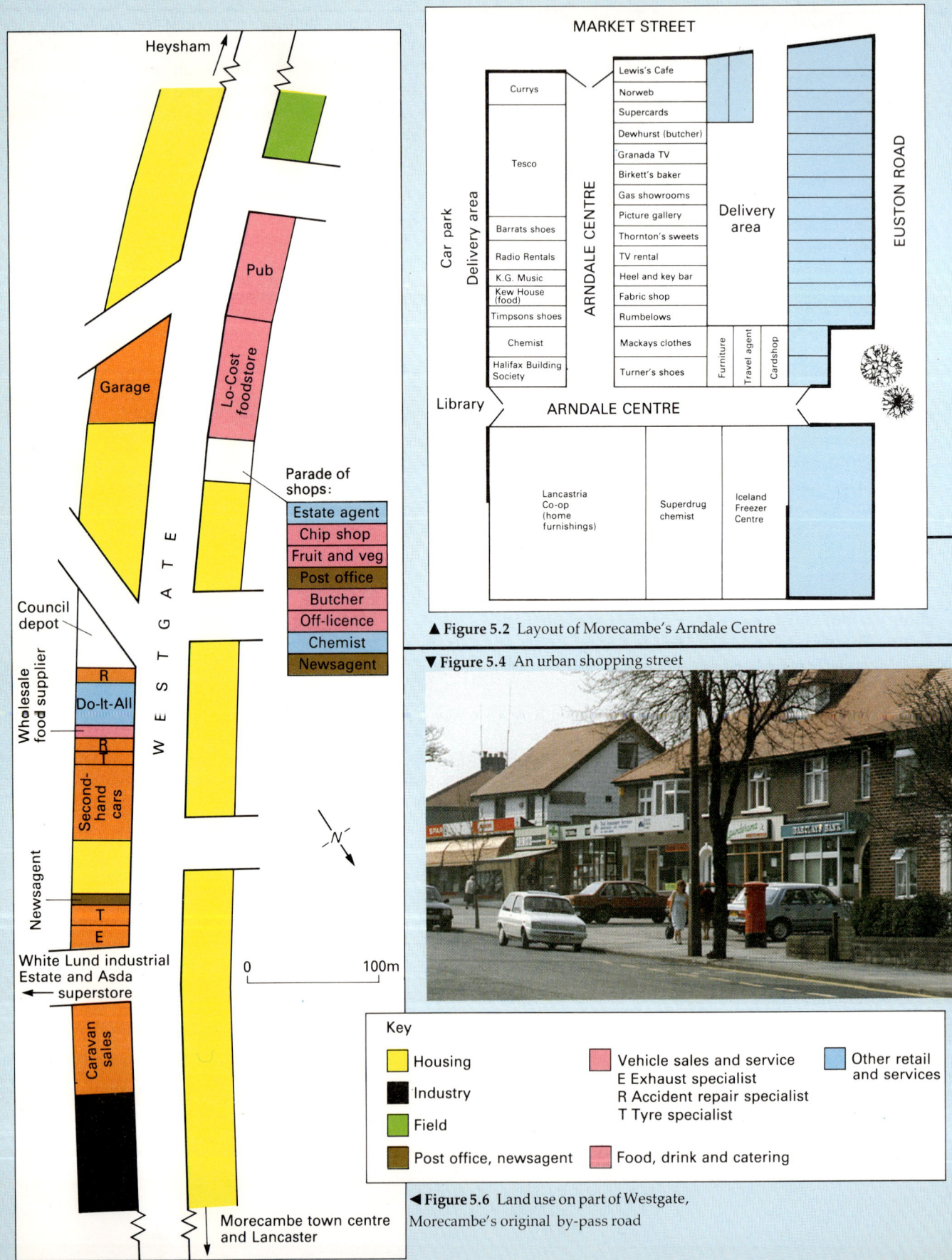

Figure 5.2 Layout of Morecambe's Arndale Centre

MARKET STREET

Lewis's Cafe
Norweb
Supercards

Currys

Car park
Delivery area

Tesco

Barrats shoes
Radio Rentals
K.G. Music
Kew House (food)
Timpsons shoes
Chemist
Halifax Building Society

ARNDALE CENTRE

Dewhurst (butcher)
Granada TV
Birkett's baker
Gas showrooms
Picture gallery
Thornton's sweets
TV rental
Heel and key bar
Fabric shop
Rumbelows
Mackays clothes
Turner's shoes

Delivery area

Furniture
Travel agent
Cardshop

EUSTON ROAD

Library

ARNDALE CENTRE

Lancastria Co-op (home furnishings)

Superdrug chemist

Iceland Freezer Centre

▼ **Figure 5.4** An urban shopping street

Figure 5.6 Land use on part of Westgate, Morecambe's original by-pass road

Heysham

W E S T G A T E

Pub

Garage

Lo-Cost foodstore

Parade of shops:
Estate agent
Chip shop
Fruit and veg
Post office
Butcher
Off-licence
Chemist
Newsagent

Council depot

Wholesale food supplier

R
Do-It-All
R

Second-hand cars

Newsagent

T
E

White Lund industrial Estate and Asda superstore

Caravan sales

0   100m

N

Morecambe town centre and Lancaster

Key

Housing

Industry

Field

Post office, newsagent

Vehicle sales and service
E Exhaust specialist
R Accident repair specialist
T Tyre specialist

Food, drink and catering

Other retail and services

Type of business	No. of units
Bank	3
Building Society/Estate Agent	3
Clothing/wool	5
Food/drink	13
Garage/bicycle accessories	3
Hairdresser	2
Hardware/electrical	2
Medical	2
Newsagent	3
Post office	1
Others	4
Total	41

▲ **Figure 5.3** Shop-types in Bare's suburban shopping centre

▶ **Figure 5.5** A corner shop

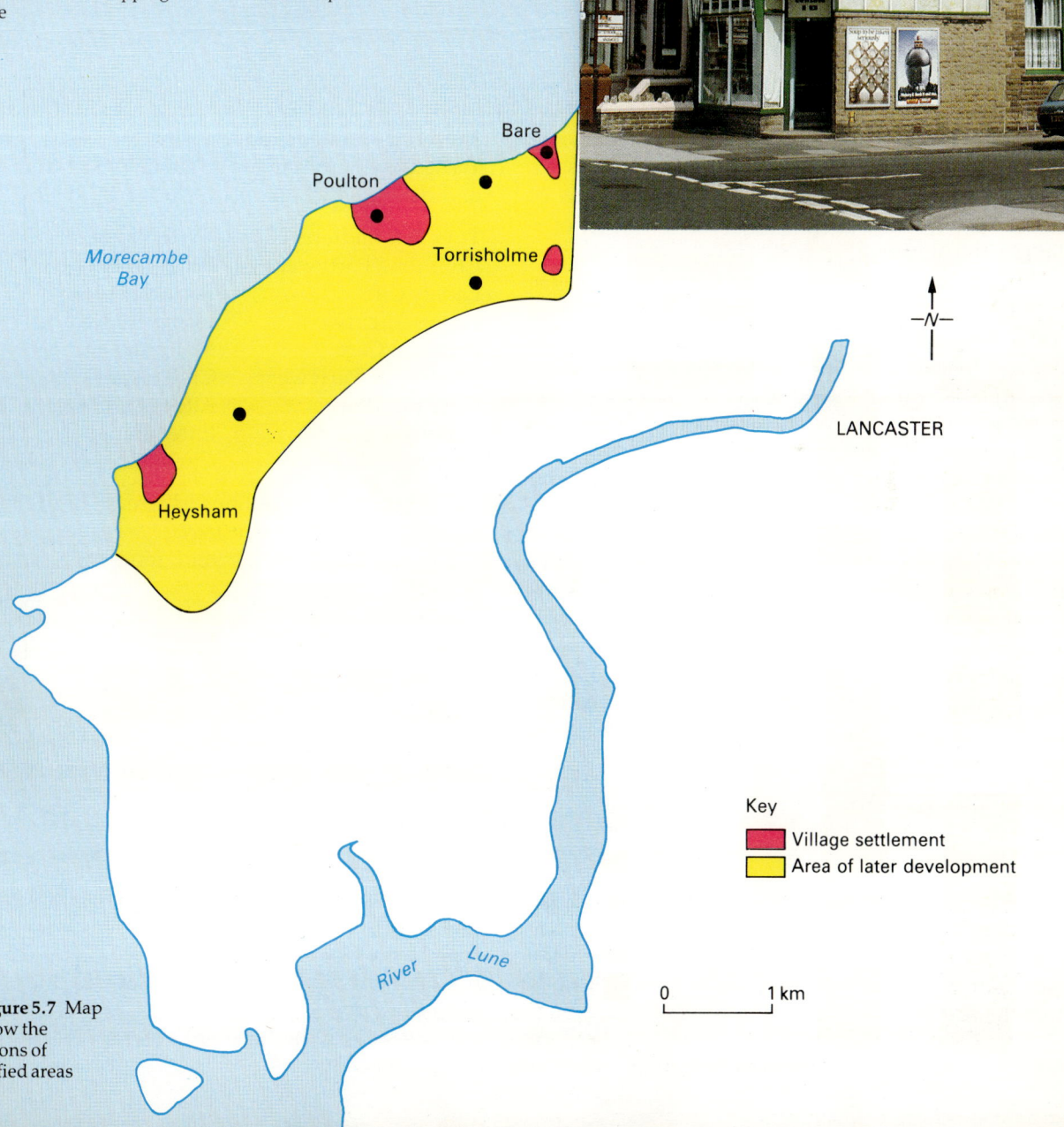

Key
  ▮ Village settlement
  ▮ Area of later development

▶ **Figure 5.7** Map to show the locations of specified areas

5.3 *Shopping trends*

▶Figure 5.8

In recent years there has been a move towards fewer and larger shops, particularly the superstores owned by groups such as Asda. Two-thirds of all food-sales are now made by these superstores, whilst tens of thousands of corner shops have ceased trading over the last thirty years (Fig. 5.8).

The stores have become more important because they can offer a much wider choice of goods – at prices which the small corner shops cannot match. Each group has a huge annual turnover and has to buy 'in-bulk' (in *very* large quantities). This allows them to buy at a lower price, and pass on this saving to the customer. Some have their own pumps selling cut-price petrol; they do this to attract car owners, who often buy most of what they need for a whole week *or month* in one shopping trip. Most stay open late on Thursday and Friday evenings to allow working people to avoid shopping at the week-end. They also accept payment by credit cards.

All kinds of food shops now sell a wide range of **convenience foods**. These are pre-packaged to make them last longer and easier to prepare for eating. Most people now have their own refrigerators and many have freezers.

1 What changes have taken place recently in:
   a The number of shops?
   b The size of shops?

2 **a** Why are supermarkets so keen to attract car owners?
   b What measures have they taken to attract the custom of car owners? (Try to add to those described in the text.)

3 Study this contents list for a recent Christmas shopping guide produced by the Asda group. Write down any ranges of goods (apart from food and drink) which are *not* included in this list.

Toys & Childrens Gifts	TV & Video
Childrens Partywear	Personal Audio
Ladies Gifts	Portable Audio
Ladies Nightwear	Home Audio
Personal Care	Motor & DIY
Lingerie & Partywear	Christmas Cooking
Gifts for Men	Kitchen Gifts
Accessories	Entertaining
Stationery & Photo	Christmas Decorations

4 Put this group of questions to at least 10 adults *who regularly shop at a supermarket*:
   a Do you buy *most* of your food from a supermarket?
   b What other types of goods do you *regularly* buy from a supermarket?
   c In your opinion, are supermarkets:

   ☐ able to provide a wide range of goods?
   ☐ able to make your shopping quicker to do?
   ☐ able to offer lower prices than other kinds of shop?
   ☐ able to make shopping more pleasant?
   ☐ 'friendly' places in which to shop?

5 **a** Use a variety of techniques (graphs, etc.) to display the information gained from your questionaire-based interviews.
   b Comment on the information you have displayed.

5.4 Shopping facilities in Runcorn New Town

The building of a new town at Runcorn was approved in 1964. Unlike most of the earlier new towns, the one at Runcorn was built next to an existing settlement (Fig. 5.9). Runcorn *Old* Town then had a population of 28 500 and all the usual shopping facilities of a town that size, including a number of chain stores such as Boots and Woolworths.

It was planned for the combined populations of Runcorn Old and New Towns to reach 100 000 by the end of this century. Most of the newcomers would be parents aged 25 to 50 from Liverpool and Manchester, with children of school or pre-school age. It was considered very important to design shopping facilities within the new town which would meet the needs of these families. That is why it was decided to build some shops in every large housing area as well as in the town centre.

The parade of shops in Halton Brook is typical of those located in these housing areas (or **neighbourhood units**, as they are often called). Figure 5.10 shows that the shops there can provide most day-to-day needs but do not offer luxury goods. To obtain these, shoppers have to travel into the town centre.

The town centre complex is called Shopping City (Fig. 5.11), and it has all the main services needed by a sizeable town. In addition to covered parking for 2 400 cars there are law courts, a police station, a library and a bus station. The building, which was opened in 1971, also has 11 230 m^2 of office accommodation for use by private companies as well as the local council. Cinemas, bars and restaurants give the people an opportunity to relax and meet socially. A boiler house nearby provides heating for the whole building as well as Southgate residential estate whose 6 000 people are linked to Shopping City by an overhead walkway.

◄ **Figure 5.9** Runcorn Old Town from the air

▲ **Figure 5.10 (top)** Neighbourhood shops, Halton Brook, Runcorn New Town

▲ **Figure 5.11 (above)** Town centre Shopping City, Runcorn New Town

Figure 5.12 shows the layout of the traffic-free shopping deck. The 46 000 m² of shop floor space have been divided up into 120 units – one of them large enough to accommodate a major Tesco supermarket. The central feature of Shopping City is the Town Square, pictured in Fig. 5.13. This is a popular meeting place and has fitted carpets, seating areas, an ice cream kiosk and even a toddlers' roundabout! The major main stores, a restaurant and a public house are grouped around the Square. Four wide **shopping malls** lead from it and give easy access to all the other facilities. White terrazo-tiled floors and marble-lined walls provide a pleasant shopping environment which is warm, well-lit and very easy to keep clean.

1 In what ways are Runcorn New Town's shopping facilities different to those in most older British towns?

2 a Do you think that the new town has been successful in meeting the shopping needs of young families?
  b Give the reasons for your answer to a.

3 a Which residential estate is nearest to Shopping City?
  b How many people live on this estate?
  c How is the estate linked to Shopping City?
  d How do people living on the *other* estates get to Shopping City?

4 Make a list of the services and facilities provided by Shopping City, *not* including its shops.

5 a State briefly whether *you* would like to do your weekly shopping in Shopping City.

  b Give as many reasons as you can for your decision.

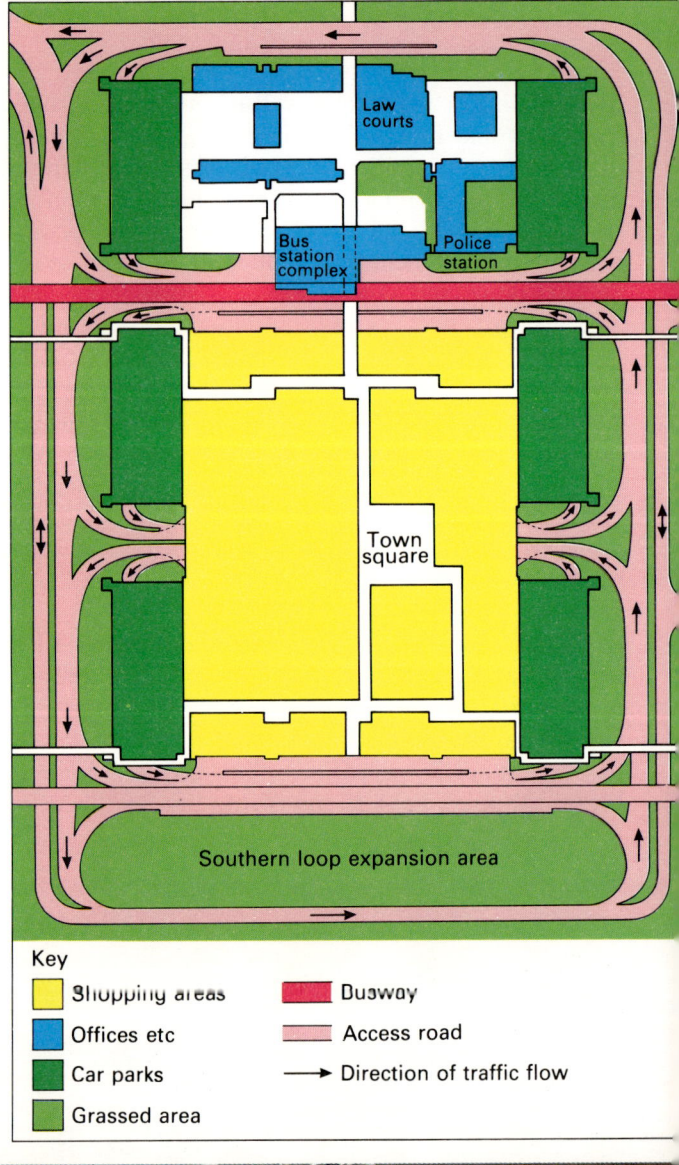

Key

🟨	Shopping areas	🟥	Busway
🟦	Offices etc	🟪	Access road
🟩	Car parks	→	Direction of traffic flow
🟩	Grassed area		

▶ **Figure 5.12 (top)**
Layout of Runcorn New Town's Shopping City

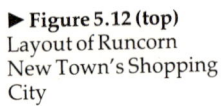

▶ **Figure 5.13 (right)**
The Town Square, Shopping City

5.5 Shopping facilities abroad

You should now know that shops reflect people's life-styles very closely indeed. The two case studies in this unit provide extra evidence that this is true.

Stacking stalls in Calcutta

Unit 3.1 introduced you to Calcutta and will have improved your perception of this conurbation in India. If you had read that unit carefully, you will know that Calcutta is a city of great contrasts; while most of its people would be regarded as very poor in a developed country, there are others who are wealthy by *any* standards. Calcutta's shopping facili-ties have, therefore, developed to meet the needs of both groups of people. The stores used by the rich stock every possible kind of luxury – at prices to match (Fig. 5.14).

At the other extreme are **stacking stalls** (Fig. 5.15). In this example, the ground floor stall-holder is sell-ing cups of tea and small portions of deep-fried sweets. His fellow trader on the 'upper storey' offers betel nuts sprinkled with spices and wrapped in leaves. These stalls cater for those who can only afford to buy in small quantities. Trade is brisk, how-ever, and some stacked stalls near to the one in this picture are *three* storeys high!

▲ **Figure 5.14** Luxury dress shop in Calcutta – such shops are to be found in *all* large Third World cities

▶ **Figure 5.15** Stacking stalls in central Calcutta

A regional shopping centre in North America

This second case study is of a **hypermarket** (the word is taken from the French *hypermarche*). Hypermarkets are much larger versions of the British supermarket described earlier; they also offer a much wider range of goods and services. Their aim is to provide 'one-stop' shopping in which clients can obtain – quite literally under one roof – almost everything they are likely to need.

In the example shown in Fig. 5.16 two major department stores act as 'magnets', attracting large numbers of people on to the site. They also create trade for the smaller businesses clustered around them. These include chemists, banks, beauty salons and insurance offices – all of them newcomers to the British supermarket scene. Both large and small units are grouped around a shopping mall in which exhibitions and displays are regularly held. The mall is centrally-heated in winter and air-conditioned in summer for the comfort of shoppers, and has many access points from the huge car park which surrounds the complex.

Hypermarkets have been much slower to appear in Britain than in North America and certain European countries. This is partly because they are **land intensive** (need large amounts of land) and are located *outside* the built-up area. As Units 8.4 and 8.5 will explain, these areas are often protected against large-scale developments.

Another reason is the hostility of long-established traders within the CBD. They *know* that hypermarkets will undercut their prices, take trade away from them, and reduce their profits so much that they may have to close down. Planning permission for the hypermarket type of complex has therefore been very hard to obtain in Britain. In spite of this, it seems certain that we shall be following the American example in the not too distant future.

1 Explain what a 'stacking stall' is, *and* how it meets the needs of Calcutta's poor.

2 **a** What is meant by the expression 'one-stop shopping'?
b Describe the ideal type of site for a hypermarket.
c Why have very few hypermarkets been built in Britain so far?

3 Describe how a family (with teenage boys and girls) might spend a useful *and enjoyable* day at the hypermarkets shown in Figs 5.16 and 5.17. Their 'day' could include some social activities in the evening.

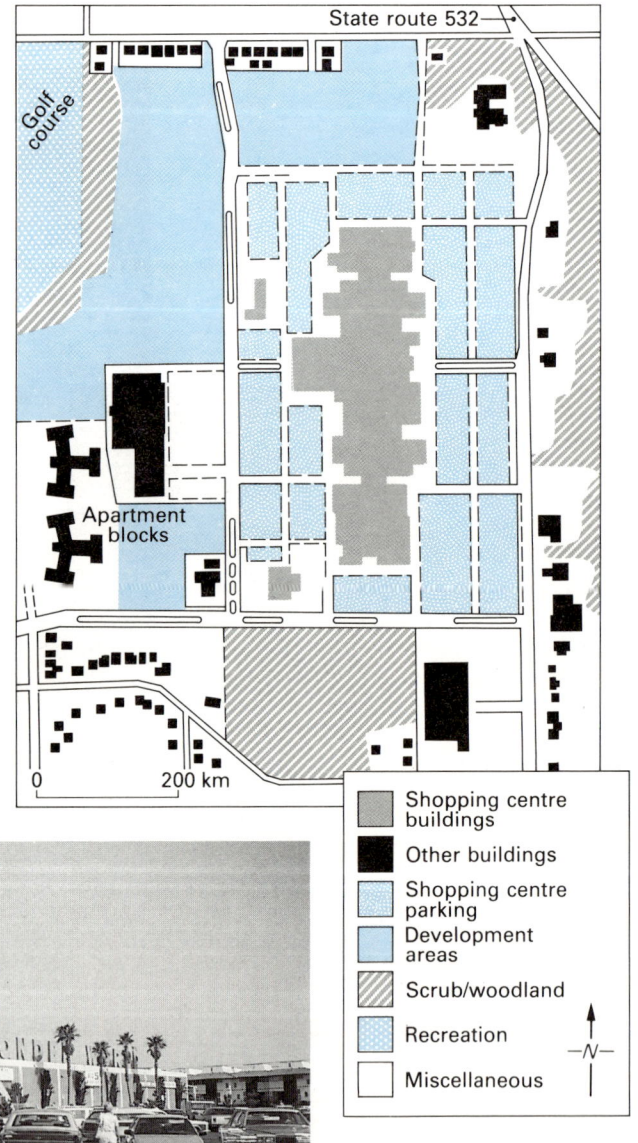

■ (dark grey)	Shopping centre buildings
■ (black)	Other buildings
▨ (light blue)	Shopping centre parking
▨	Development areas
▨	Scrub/woodland
▨	Recreation
□	Miscellaneous

▲ **Figure 5.16** Layout of the Chapel Hill hypermarket at Akron, Ohio

◀ **Figure 5.17** View of the American hypermarket Wonder World in Las Vegas, USA

6

Other Facilities

6.1 Introduction: standards of health

Chapter 5 described how shopping facilities vary according to the wealth – and the population – of an area. The units in this part do the same for three other aspects of everyday life.

Turn back to the table on page 8, which lists the *life expectancies*, *infant mortality rates* and *daily intake of calories* for some typical developed and developing countries. These figures show all too clearly that, while some countries enjoy generally high standards of health, others do not.

These figures also suggest that there is a strong link between bodily health and the quantity of food eaten. This is certainly true, but the relationship isn't quite as simple as it might appear. It is known that over-eating tends to make us over-weight; also that it increases the risk of suffering from heart trouble. Luckily, most of us can reduce our chances of heart failure by exercising some self-control.

The situation facing the *majority* of people in the Third World countries is somewhat different. Many of them are under-fed, and have very limited opportunities of increasing their intake of food to a satisfactory level. It is known that well over half of all the people in these countries suffer from **under-nourishment** – a lack of sufficient *quantity* of food. This leads to loss of weight, lethargy (lack of energy) and eventually starvation (Fig. 6.1). Five reasons for this situation are listed below:

- While many Third World farmers have made great improvements to their land, others lack the technology (the knowledge and know-how) to obtain maximum food yields from the soil.
- They also lack the finance – and confidence – to buy fertilisers and improved types of seed. These could give them much higher crop yields.
- Some of the food they produce is eaten by rats and cockroaches. Better storage facilities would reduce this loss.
- Many developing countries grow *cash crops* (to export for profit) on their best land – land which *could* be used to grow more food for the people at

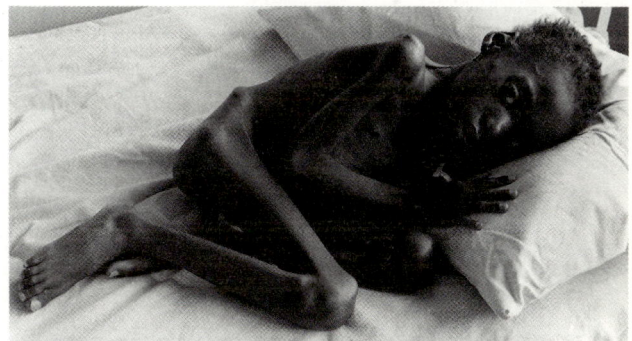

▲ **Figure 6.1** This child is suffering from severe under-nourishment and has not had *enough* to eat for many months

home. Cash crops first became important to them after they were colonised by Europeans, and this pattern of farming has continued ever since.
- These countries are generally too poor to import extra food from others which have a surplus.

Millions of people in the Third World also suffer from **malnutrition**. This means they do not receive all the different *types* of food needed to keep them healthy. Figure 6.3–6.6 illustrate some effects of not getting enough protein and vitamins. Equally worrying is the fact that malnutrition reduces the body's resistence to infection. Diseases such as cholera, yellow fever and malaria are quite common anyway where there is unsafe drinking water and poor sanitation (toilet and waste-disposal facilities). Inadequate medical facilities are another reason for widespread illness in the Third World. The 'cycle' in Fig. 6.2 shows how poor feeding can affect countries which have to grow most of the food they need.

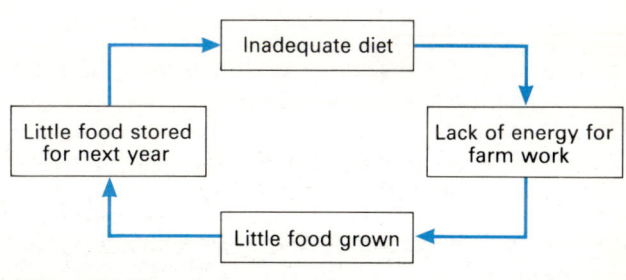

▲ **Figure 6.2** The poor farmer's cycle

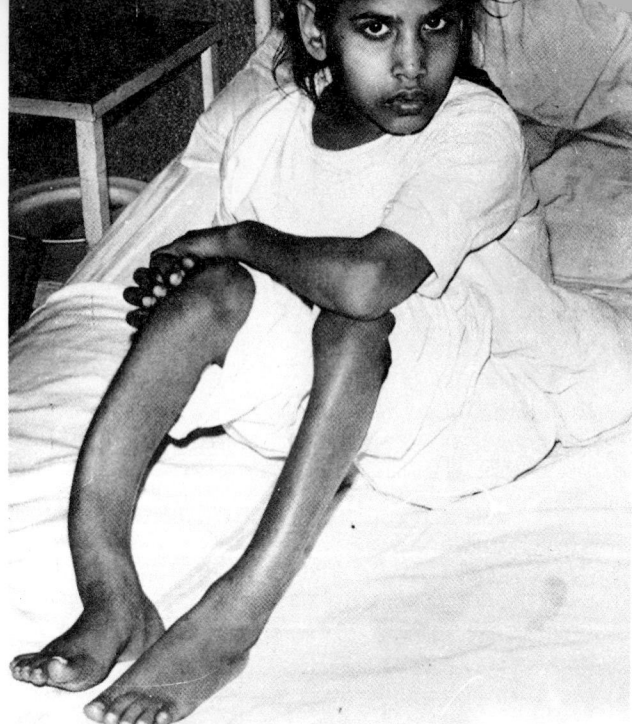

▲ **Figure 6.3** This girl suffers from *rickets*, a condition which affects the bones of children who do not receive enough Vitamin D. This can be obtained from eating butter, milk and fish

▼ **Figure 6.5** This man was not born blind. He has cataracts caused by lack of Vitamin A. Eating vegetables, fruit and fish could have saved his sight

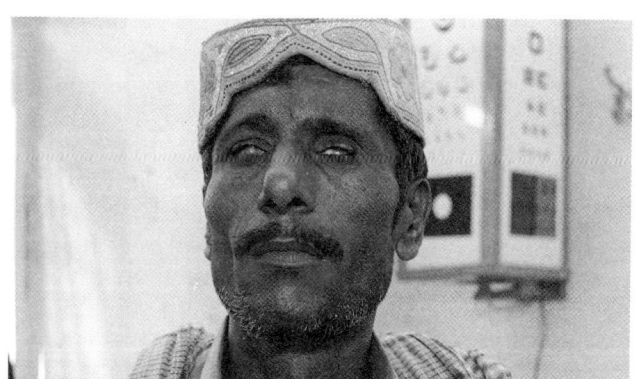

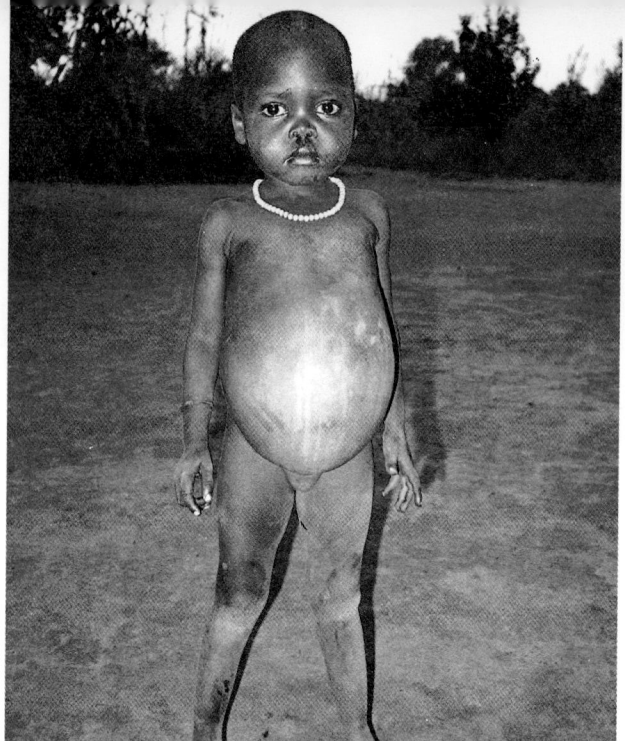

▲ **Figure 6.4** This girl is suffering from kwashiorkor. She needs a balanced diet which includes meat, fish, eggs and cheese; these are rich in protein

▼**Figure 6.6** This man has scurvy because he has not eaten enough fresh vegetables or fruit. Lime-juice would be very helpful in raising his intake of Vitamin C. Scurvy used to be very common among sailors who had to spend long periods at sea, and so couldn't eat much fresh food

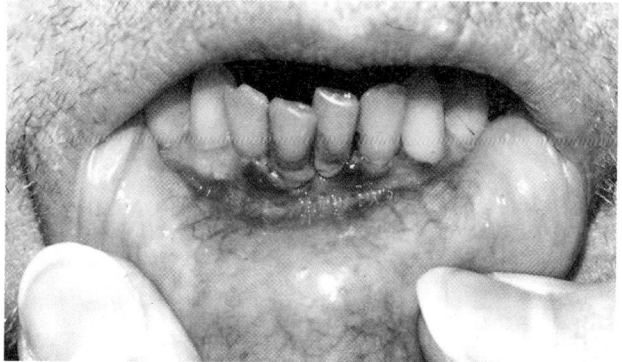

1 a Which cause of death affects many people in *developed* countries?
b After class discussion, suggest ways in which people can reduce their chances of dying from this cause.

2 a Explain the meanings of 'under-nourishment' and 'malnutrition' very carefully.
b Use the five reasons listed on page 81 to explain why many people in Third World countries do not have *enough* to eat.

3 Draw two pie graphs to show the following information:

Type of Food	Proportion of total food intake in	
	United Kingdom	**Uganda**
Carbohydrates (provide calories)	169°	317°
Vitamins	90°	32°
Protein	101°	11°

b Describe briefly what your two pie graphs show.
c With the help of Figs 6.3–6.6, describe some of the effects of the eating patterns shown by your completed graphs.

4 Study this table very carefully, then use full sentences to answer the following questions based on it.

Country	Per capita GNP in US $	Number of doctors per 10 000 people	Daily intake of calories (as a % of basic need)
Bangladesh	140	0.7	84
Belgium	11 920	20.3	160
India	260	2.5	87
Jamaica	1 180	2.8	119
Japan	10 080	5.4	124
Nigeria	870	0.6	91
Philippines	790	3.2	116
Spain	5 640	18.3	135
Sri Lanka	300	1.7	102
United Kingdom	9 110	13.1	132

a Which is the richest country?
b Which is the poorest country?
c Which country appears to have the best medical facilities?
d Which country appears to have the poorest medical facilities?
e Which country has the highest daily intake of calories?
f In which countries do many people not have enough calories in their diet?
g In which country do most people have just about the right number of calories in their diet?

5 a Copy out the *first three* columns of the table in Question 4, than add another four columns to the right of this table; the extra columns should have these headings:

e Complete the last column. Belgium should have 0 (0 × 0 = 0).
f Write down the total of all the numbers in the last column.
g Use this formula to find the number which can tell us how closely these two sets of information are matched.

$$1 - \frac{6 \times \text{'the total number of squared differences'}}{(\text{'the number of rankings'})^3 - \text{'the number of rankings'}}$$

$$= 1 - \frac{6 \times \text{the number you found in (f)}}{(10 \times 10 \times 10) - 10}$$

You can work out the rest for yourself! The number you get is called the **Spearman Rank Correlation Coefficient**, named after the man called Spearman who invented it.

h Write down the number you have worked out – to two decimal places. (Hint: the correct answer is one of the following: 0.35; 0.44; 0.57; 0.61; 0.79; 0.83; 0.98.)

i Finally, write down the description of the relationship which matches your number. The four possible descriptions are:

'A very strong link' (for numbers 0.70–1.0)
'A fairly strong link' (for 0.40–0.69)
'A weak link' (for 0.20–0.39)
'Almost no link' (for 0.00–0.19).

6 a Repeat Question 5 for *per capita GNP* and *daily intake of calories*.
b How does this second relationship compare with that for Question 5?
c What does this tell you about the usefulness of GNP as an indicator of the overall quality of life in a country?

Ranking of countries in order of per capita GNP	Ranking of countries in order of doctors per 10 000 people	Differences in ranking	Squared differences in ranking

b Complete the first new column by writing 1 for Belgium (which has the highest per capita GNP), 2 for Japan, and so on.
c Complete the second new column by writing 1 for Belgium, 2 for Spain, and so on.
d Complete the next column by writing the differences between the pairs of numbers you have just entered for b and c. Belgium will have 0 (1 − 1 = 0). There is no need to use the minus sign (e.g. 2 − 4 = 2, not − 2).

6.2 *Education and literacy*

Britain spends billions of pounds every year maintaining our schools, colleges and universities, paying teachers and lecturers, and buying books and equipment. We would like to spend much more than we do, but the government insists that times are hard and education is only one of many commitments they have to meet. Most Third World countries are permanently in a state of hardship and can afford only a fraction of what we spend on each pupil. Colombia, in South America, is a typical developing country in this respect. Some facts about its educational system are given below; others are shown in Figs 6.7–6.9.

□ Its schools do not provide free textbooks, exercise books or writing materials. Parents have to buy these themselves.

□ Each province (large administration district) has to maintain its own schools. Many of Colombia's provinces are remote, rural areas whose poor farmers cannot afford to pay enough tax to maintain to maintain the schools properly. The industrial provinces are better off, but many people there manage to avoid paying all the tax they should.

□ The cities attract the best teachers. Many of them were born and educated in the cities and are used to living in them. There are other reasons too. The cities have better social facilities, colleges offering advanced training courses, and more housing equipped with proper sanitation and a safe water supply.

□ Much of the teaching is carried out in private schools. 20% of all primary schools and no fewer than 55% of the country's secondary schools are privately owned. Most of these are in the cities, where there are more wealthy people who can afford to pay their fees.

□ Parents realise that the better-paid jobs go to those with the highest qualifications, so the little they can spend on their children's education isn't likely to secure them a decent job.

Clearly, most Colombian children do not have the same educational opportunities as those in the developed countries. This is reflected in the literacy rates. 99% of British adults are literate (can read and write well). The equivalent figure for Colombia is 81%. In fact, Colombia has worked very hard to achieve this figure for it was only 63% in 1960. The present average for the whole of the Third World is 68%.

Many developing countries are now taking literacy much more seriously. They realise that administration becomes easier as the literacy rate increases; they also know that literate countries stand a far better chance of becoming 'healthy, wealthy and wise'. For example, Tanzania (in East Africa) has revolutionised the educational system which it inherited on becoming an independent country. Under the inspired leadership of President Julius Nyere it has developed 'Education for Self-Reliance' for all its people, who have benefited greatly from their new-style lessons.

▲ **Figure 6.7** Tania is 12 years old. Only two-thirds of her friends of the same age have spent two years at school

▲ **Figure 6.8** Maria has been very lucky. She has been to a primary school, a secondary school (only 4% of Colombian teenagers receive a full course at secondary school) and is now at a university training to be a teacher. You can read her letter in Fig. 6.9

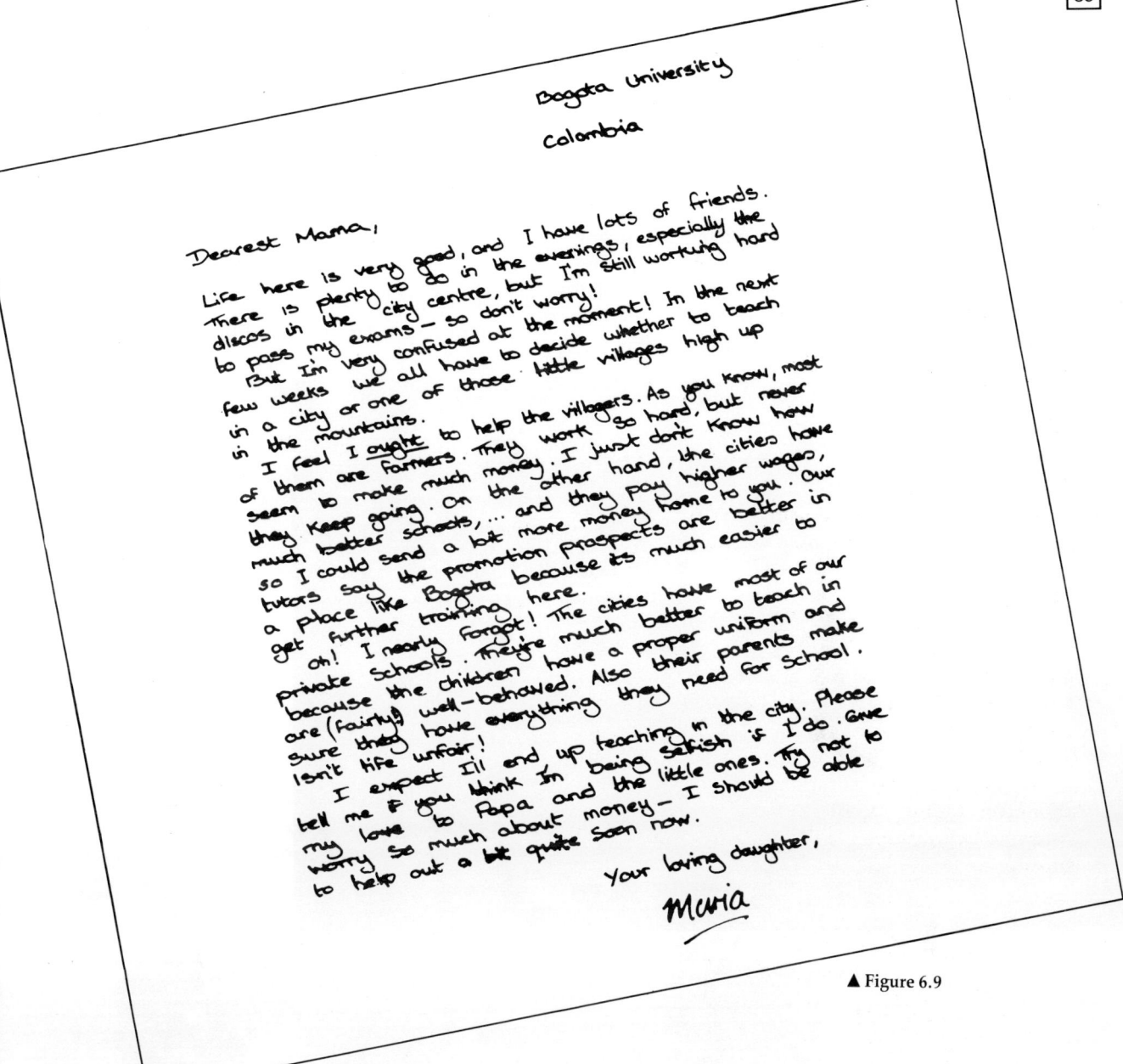

Bogota University

Colombia

Dearest Mama,

Life here is very good, and I have lots of friends. There is plenty to do in the evenings, especially the discos in the city centre, but I'm still working hard to pass my exams — so don't worry! But I'm very confused at the moment! In the next few weeks we all have to decide whether to teach in a city or one of those little villages high up in the mountains.

I feel I ought to help the villagers. As you know, most of them are farmers. They work so hard, but never seem to make much money. I just don't know how they keep going. On the other hand, the cities have much better schools, ... and they pay higher wages, so I could send a bit more money home to you. Our tutors say the promotion prospects are better in a place like Bogota because it's much easier to get further training here.

Oh! I nearly forgot! The cities have most of our private schools. They're much better to teach in because the children have a proper uniform and are (fairly) well-behaved. Also their parents make sure they have everything they need for school. Isn't life unfair!

I expect I'll end up teaching in the city. Please tell me if you think I'm being selfish if I do. Give my love to Papa and the little ones. Try not to worry so much about money — I should be able to help out a bit quite soon now.

Your loving daughter,

Maria

▲ Figure 6.9

1 Explain the meanings of the terms:

illiterate
literate
literacy rate

2 After class discussion, suggest reasons why literate countries are more likely to become:
  a 'healthier'
  b 'wealthier'
  c 'wiser'.
than countries with a much lower literacy rate.

3 Copy out each sentence, then use one of the numbers in brackets to fill its blank space.
  a In Colombia, approximately one person in every . . . is illiterate (2, 3, 4).
  b In Colombia, only two out of every . . . children spend two or more years in a primary school (2, 3, 4).
  c Since 1960, Colombia's literacy rate has increased by about one-. . . (third, quarter, fifth).

4 Read the letter in Fig. 6.9. If you were Maria's mother, what advice would *you* give your daughter on deciding where to teach? Your reply to Maria should also be written in the form of a letter.

6.3 Recreational facilities in towns

British people now spend more time and money on **recreation** than at any time in the past. Modern town planners are aware of the growing need for recreational facilities and try hard to include them in their designs for the future.

This was not always the case, however, for during the last century the building of houses, factories and railways had priority. Many large areas of Victorian terraced housing still do not have enough open space which can be used for recreation. Even so, the Vic-

torians were not completely blind to the need for *some* recreational open space, and most nineteenth-century industrial towns did have a 'Corporation Park' and a number of bowling greens and tennis courts (Fig. 6.10). Many allotments were also laid out during this period. Although not particularly attractive to look at, they did provide an opportunity for healthy exercise and making some savings on a family's food bill! Urban renewal schemes have increased the amount of open space within these older areas. Grassed areas – some of them landscaped and

▲ **Figure 6.10** Regent Park, Morecambe

▶ **Figure 6.11** New recreational open space in Preston

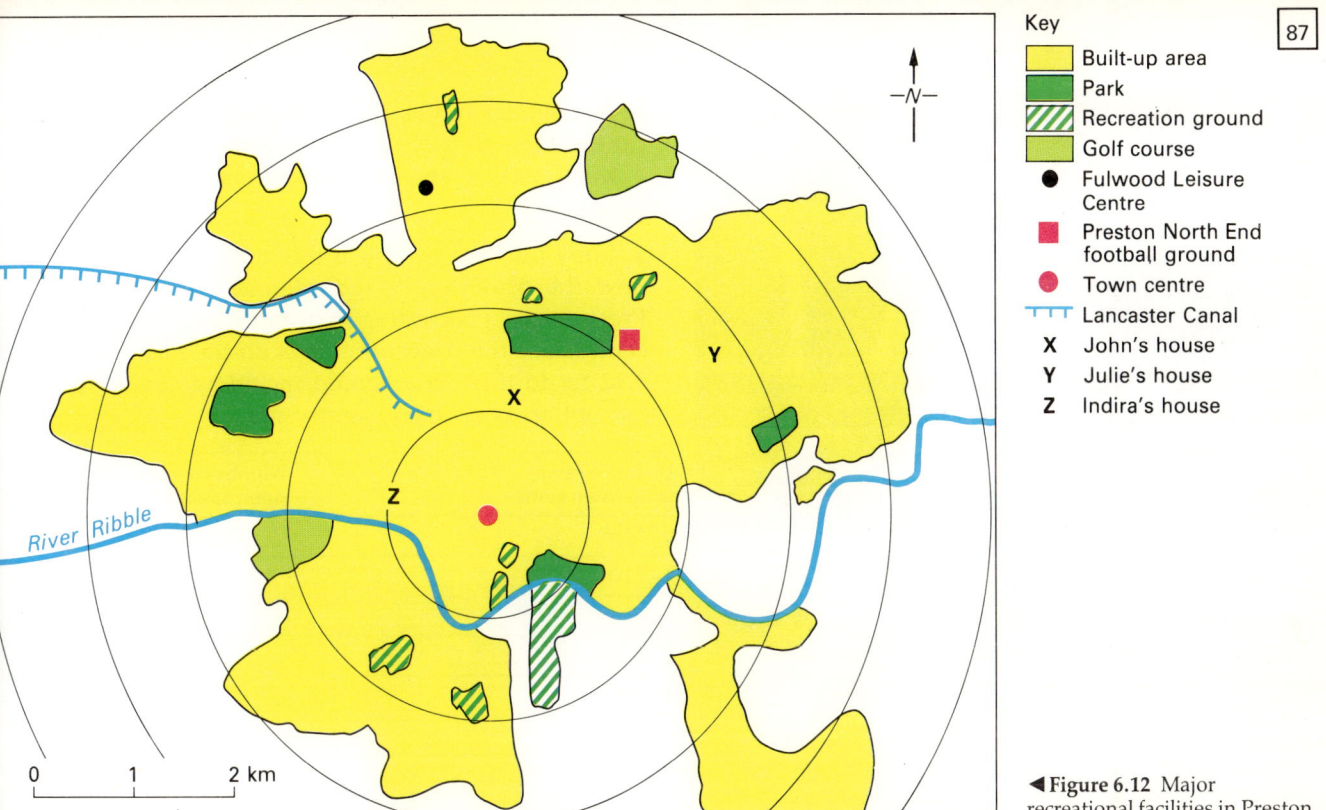

Key
- Built-up area
- Park
- Recreation ground
- Golf course
- ● Fulwood Leisure Centre
- ■ Preston North End football ground
- ● Town centre
- ━━━ Lancaster Canal
- X John's house
- Y Julie's house
- Z Indira's house

◀ **Figure 6.12** Major recreational facilities in Preston

planted with bushes and trees – make valuable extra 'breathing spaces' within the maze of residential and industrial buildings (Fig. 6.11).

The distribution of recreational open space within our urban areas tends to vary a great deal. Figure 6.12 shows that this is certainly true of Preston – a fairly typical industrial town in central Lancashire which has expanded greatly over the last 200 years. Notice that the heart of this town has very little open space, but that the Victorian areas (within 2 km of the town centre) include some allotments and two large parks. The twentieth-century suburbs seem to be less fortunate, but nearly all of their houses have private gardens.

Some ancient cities have considerably more open space in their inner areas than Preston. The reasons for this are mainly historical. The large parks on the western side of central London (Fig. 7.3 on page 92) are remnants of medieval hunting forests. These had to be within easy reach of the royal palaces and homes for the nobility, who preferred that side of the city to the more industrialised zones further east.

Our *new towns* are particularly well off for open space as their planners were instructed to give a high priority to recreational and social facilities. The planners of Basildon New Town in Essex were *very* clear about the levels of provision they hoped to achieve (see Fig. 6.13).

▼ Figure 6.13

Age range	Facility	Maximum distance from home
under 7 years	play area	50 metres
7–13 years	play area	150 metres
over 13 years	area for casual ball games	400 metres

This study would be incomplete without some discussion of *indoor* recreational facilities. Municipal swimming baths have been a common feature for over a hundred years, but increasing interest in health and physical fitness has led to the building of many other publicly – and privately – owned facilities. Well over one hundred leisure centres have now been developed by local councils, and private health clubs are definitely on the increase. Some of these are very well equipped, with squash and badminton courts, exercise suits and saunas. Snooker is extremely popular at the moment and every large town has at least one snooker and billiards hall.

Local authorities now realise that our towns cannot provide all the recreational open space needed by their inhabitants. They have therefore created **country parks** within easy travelling distance of the main urban areas. One of these new parks is described in the next unit.

▲ **Figure 6.14** 'Snooker heroes'

1 Compare the two recreational open spaces shown in Figs. 6.10 and 6.11, under these separate headings:
   a Recreational facilities.
   b Suitability for use by:

   ☐ children (up to 13 years of age)
   ☐ teenagers
   ☐ adults (aged 20–60)
   ☐ retired people.

   c General layout and appearance.

2 **a** Copy this *radial diagram* which shows the *direct* distances which John has to travel from his home (see Fig. 6.12) to

   ☐ Fulwood Leisure Centre
   ☐ Preston North End Football Ground
   ☐ the nearest park
   ☐ the nearest stretch of Lancaster Canal.

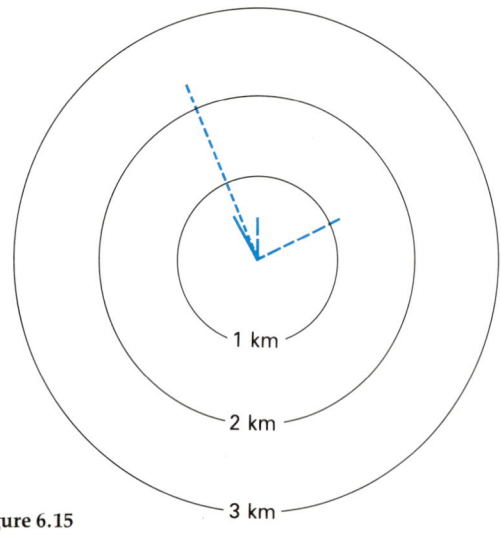

▲ **Figure 6.15**

b Plot similar radial diagrams for Julie and Indira (to the same four destinations).
c Comment on the patterns shown by the three radial diagrams.

3 **a** Copy out and complete this table. It will show how the provision of recreational open space in Preston varies according to distance from the town centre. Use the concentric rings on the map in Fig. 6.12 to do this. You should include golf courses, but not areas of open countryside.

Distance range from town centre	Number of recreational open spaces in range
0–1 km	
1–2 km	
2–3 km	
3–4 km	
4–5 km	
over 5 km	

b Say what your completed table can tell you about the *distribution* of Preston's recreational open spaces.

4 **a** Name the central London parks shown on Fig. 7.3 on page 92.
b Explain how these parks came to be located in this inner city area.

5 **a** Copy Basildon's ideals (in Fig. 6.13) for the availability of recreational open spaces.
b Ask at least twenty of your friends to pace out the shortest distance between home and the nearest recreational space which fits Basildon's ideals (i.e. three distances – one for each). Work out the class's average distance (in metres) for each type of recreational space.
c State how your averages differ from the ideals for Basildon, and suggest reasons for these differences.

6 **a** Draw a plan of a leisure centre in your local area.
b After discussion with your friends, suggest ways in which its facilities could be improved.

7 With the help of a street map, show how an urban redevelopment scheme has increased the amount of recreational open space in your local area.

6.4 Beacon Fell – Lancashire's first country park

The previous unit examined recreational facilities in a typical urban environment. Country parks encourage relaxation in a *rural* setting. Britain has over 200 hundred such parks, many of them within easy reach of a large town or city. Some cater for highly organised activities such as golf, sailing and horse riding. Others – including the one on Beacon Fell – cover quite a small area and offer only basic tourist facilities.

Beacon Fell is a hump-backed hill 266 metres high, lying 13 km north-east of Preston (Fig. 6.16). The hill was shaped and rounded during the Ice Age and rises quite steeply on all sides. It has a long and interesting history. The Romans built a road across its south-western flank to link their forts at Lancaster and Ribchester. Later, the Vikings herded their animals on to the fell for summer grazing.

Its commanding position overlooking a fertile plain made the fell an ideal site for a beacon – hence its name. Fires were lit there to warn local people of great danger or pass on very important news. Sightings of the Spanish Armada in 1588 were passed on in this way, as were reports of frequent invasions by Scottish armies.

In 1930 the fell became the water-gathering ground for a small reservoir. About 50 hectares of fellside were planted with coniferous trees (Fig. 6.17) as they slow down the flow of rainwater to lower ground. This helps to filter the water and prevent it from becoming discoloured.

Lancashire County Council bought all 222 hectares of the fell in 1960, and work began immediately to prepare the site for recreational use. This involved:

☐ Burning down some of the coniferous trees to improve the views from the fell.
☐ Clearing small areas of forest for car parks and picnic areas. These are widely spaced apart (Fig. 6.18) so that many people can 'lose' themselves on the fell; this avoids overcrowding and makes the best possible use of all the land in the park.
☐ Planting some deciduous trees to break the monotony of the coniferous plantation. Some of these newer trees help to screen the car parks and picnic areas from view.
☐ Improving and extending the existing road around the fell. A one-way traffic system was introduced for safety reasons; it has also avoided the need to widen the road or have passing-places at intervals.
☐ Creating a small lake to store water in case of fire.

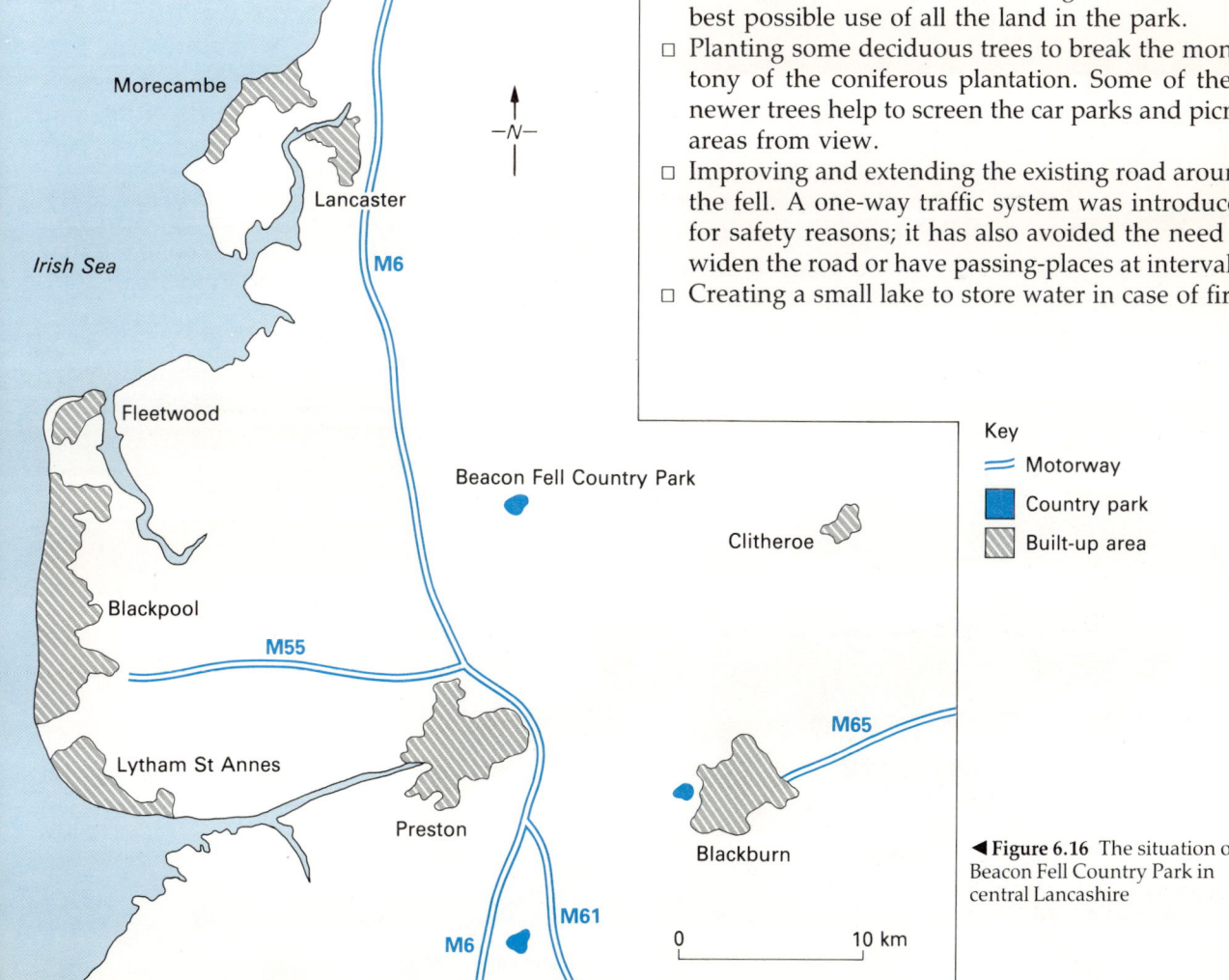

Key
═ Motorway
▮ Country park
▨ Built-up area

◀ **Figure 6.16** The situation of Beacon Fell Country Park in central Lancashire

▲ Figure 6.17   ▼ Figure 6.18

Beacon Fell Country Park was opened to the public in 1970 and a Park Ranger appointed to supervise it. Toilets, a cafe and a small Information Centre were built four years later (Fig. 6.19). A second set of toilets was erected in 1984. Access to the park is free. It was expected that most visitors would come from the nearby towns of Preston, Blackburn and Lancaster, and as many as 10 000 people have been recorded on the fell at peak periods.

The main attractions of the park are its peaceful 'natural' surroundings, and the fine views on all sides. Visitors can roam at will over the whole site and there are no gates or opening-hours to restrict access. The variety of trees includes pine, larch, spruce (all coniferous), and ash, oak and beech (which are deciduous). Weasels, stoats and rabbits breed well and the bird-life is plentiful.

1 Describe the location of Beacon Fell Country Park relative to the three towns named in the text. This is best done by quoting the compass direction and distance of each town *from* the park.

2 Write a simple 'Diary of Events' in the history of Beacon Hill *and* its park. Begin each entry with the year in which the event took place. (Hints:

☐ The Roman road was built about AD 78.
☐ The Vikings began to settle in the area during the ninth century.
☐ The main Scottish invasions took place in 1322 and 1389.)

3 Note down what had to be done to convert Beacon Fell into a country park.

4 Describe the ways in which Beacon Fell Country Park can meet the recreational needs of the local urban population. (Hint: Use the plan of the park in Fig. 6.20 as well as the text.)

5 Prepare a similar study of a country park in your own area. It would be interesting to point out how it differs from the one described in this unit.

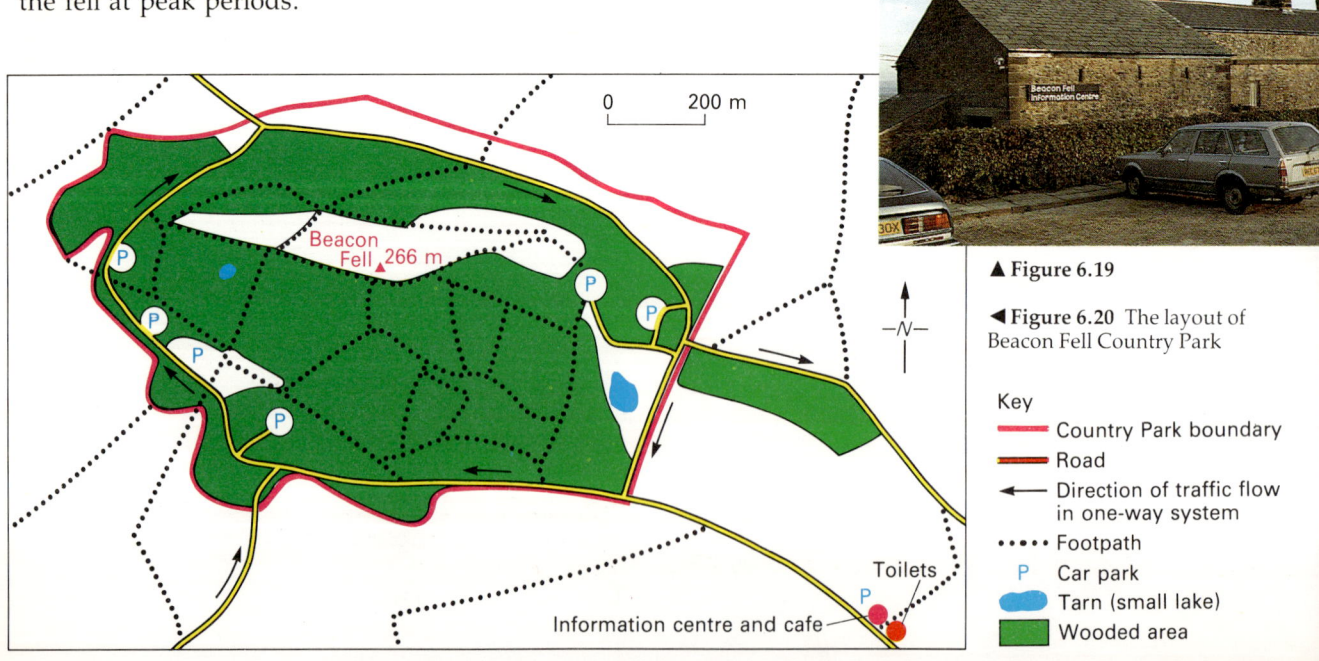

▲ Figure 6.19

◄ Figure 6.20 The layout of Beacon Fell Country Park

Key
━━━ Country Park boundary
━━━ Road
◄— Direction of traffic flow in one-way system
•••• Footpath
Ⓟ Car park
⬤ Tarn (small lake)
🟩 Wooded area

Beacon Fell ▲ 266 m

Toilets
P
Information centre and cafe

0   200 m

—N—

7

Transport Networks

7.1 Introduction: transport problems

Rush hour peak travelling times occur in the early morning and late afternoon/early evening of every working day (Fig. 7.1). They are chiefly due to commuting, but many other types of 'traffic' add to the congestion on our roads. People use them regularly to go shopping, get to school and visit friends and relatives. Recreation and tourism generate more traffic every year. Eighty per cent of all freight movements in Britain are now made by road, and the increasing size and number of heavy vehicles places a great burden on our road network. This unit considers the main problems created by heavy traffic and some of the ways in which we can reduce them.

Traffic congestion

This occurs when traffic cannot move at the legal maximum speed considered safe for a particular stretch of road. In Britain, these limits are 70 mph (112 kph) on motorways and roads in open country, and 30 mph (48 kph) on most roads in urban areas. Rush hour traffic is the main source of congestion, but Saturday shopping and seasonal holiday movements also create peaks of congestion. Localised con-

◀ **Figure 7.1** Congestion occurs when traffic cannot move freely, and is worst at rush hour periods

▲ **Figure 7.2** Traffic congestion at the London end of the M40

gestion (within a fairly small area) is often caused by special attractions such as football matches.

The motorways built in Britain since the late 1950s have certainly helped to reduce congestion along the most heavily used routes. They have allowed 'through-traffic' to avoid town centres, speeded up long-distance travel and provided safer movement for large vehicles such as coaches and juggernauts. They do however tend to *create* congestion wherever they end in built-up areas (Fig. 7.2). Many towns are now by-passed by other types of main road (see case study in next unit).

92

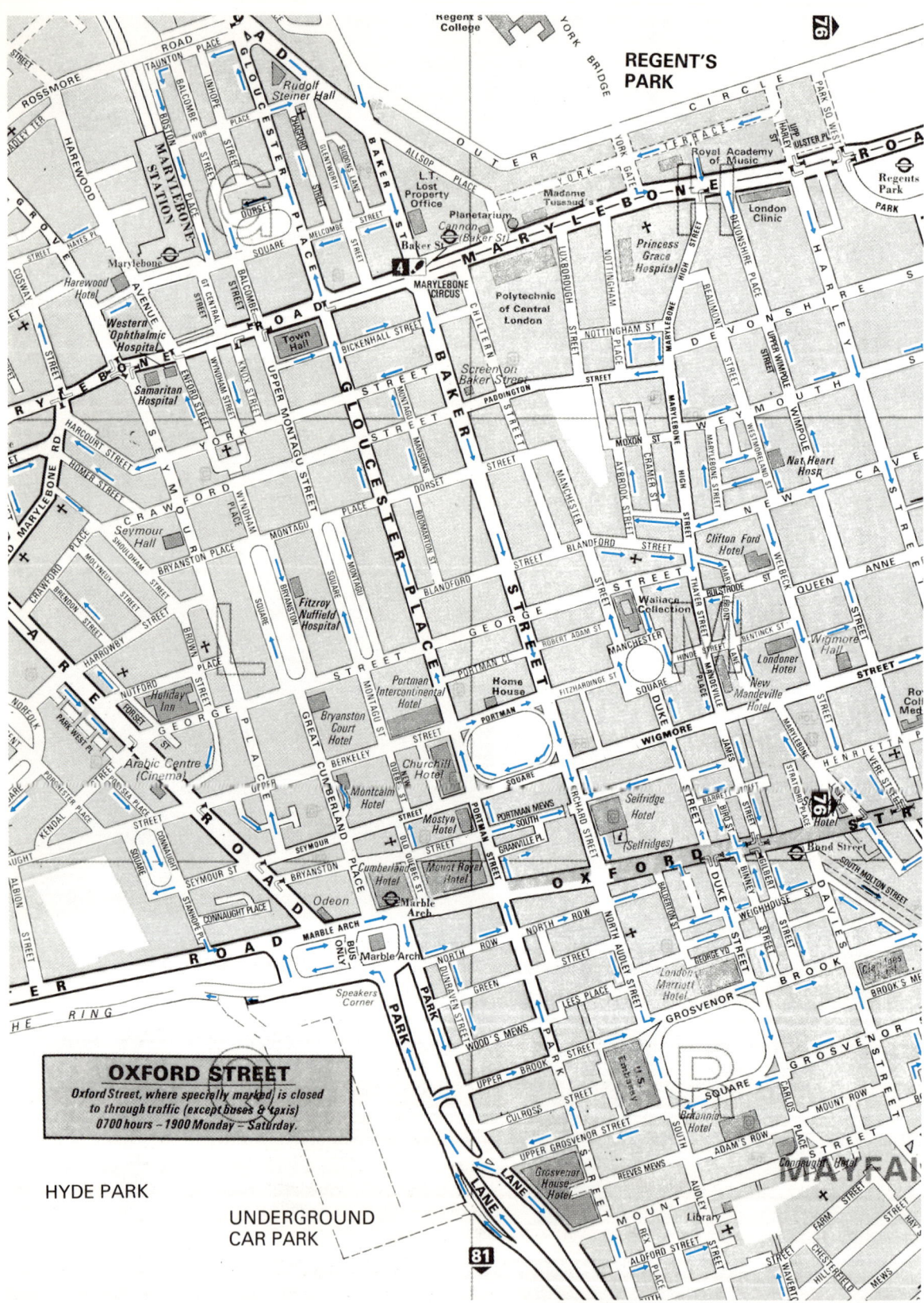

▲ **Figure** 7.3   Scale = 7 inches to 1 mile (11 cm to 1 km)

Much has also been done to speed up the flow of traffic *within* our towns and cities. Some particularly busy roads leading into them have been classified as 'urban clearways', and parking is forbidden along them at all times. Road widening is another popular way of increasing the capacity of major routes. Selected roads may have 'bus-lanes' or an ingenious system by which the direction of traffic flow along a central lane is *reversed* during weekdays. This means it can be used by incoming traffic during the morning rush hour, as well as outgoing traffic later in the day! Many offices and factories now operate a 'flexitime' system which allows workers to travel outside the two peak periods. The map of part of central London in Fig. 7.3 shows some other ways of reducing traffic congestion.

Parking problems

Vehicles can create problems while parked as well as on the move! Commercial vehicles such as buses cause congestion because of their *size*. Cars are a somewhat different matter; it is their *numbers* which most hinder the flow of traffic. If their parking were not controlled in some way, they would quickly bring traffic in central areas to a stand-still.

Planners often think of long and short stay parking as separate problems. Long stay facilities are needed for commuters, whose cars may be parked for up to ten hours every working day. Street-parking is totally inadequate for this length of time and land has to be set aside for open-air or multi-storey car parks (Fig. 7.4). Unfortunately, this type of parking does nothing to reduce the numbers of vehicles *entering* the central area. A more attractive solution is the 'park and ride' scheme by which motorists leave their cars in open-air car parks. These are sited near to their bus and railway connections into the centre.

Short stay parking is still possible on many streets, but motorists are being encouraged to use car parks instead. Street parking on weekdays is often limited by painting yellow lines by the kerb; parking meters allow drivers to park as long as they like – provided they can afford it! Traffic wardens and the police share the task of ensuring that parking restrictions are followed.

Danger and inconvenience

High volumes of traffic increase the risk of accidents between all groups of road users. Dividing wide roads into clearly marked traffic lanes has helped to reduce the number of serious accidents; so has segregating traffic flowing in opposite directions. Cycleways are another helpful measure, but they are usually found outside central areas. Can you suggest reasons why?

Pedestrians need protection too! 'Pelican' crossings controlled by traffic lights and school crossings supervised by 'lollipop' men and women are common sights on our roads. Busy shopping areas may be made much safer – and more pleasant – by turning them into 'pedestrian precincts'. Our new towns have taken both vehicle and pedestrian safety very seriously indeed, and some of the most recent ones segregate the two completely.

◀ **Figure 7.4** Multi-storey car park at Preston bus station

Environmental problems

Heavy traffic can affect our quality of life in a number of ways. Exhaust fumes contain lead and carbon monoxide, both of which are dangerous to health, as well as dirt which settles on buildings and goods in the shops. It can also make breathing quite unpleasant. These hazards, together with the vibration and constant noise produced by heavy traffic, often reduce the value of housing near to motorways and other major routes.

1 a What is a 'rush hour'?
  b When do the rush hours take place on a typical working day?

2 Describe the main causes of traffic congestion on Britain's roads.

3 Write out these facts about motorways under two headings: Advantages of motorways
                                 Disadvantages of motorways

Vehicles can travel very fast.
Most motorway accidents are very serious.
There are fewer accidents on motorways.
Motorways cut many farms in two.
Motorways can take large loads more safely than other kinds of road.
Motorways use a lot of land, much of it fertile lowland.
There may be traffic congestion where motorways end in built-up areas.

4 List the various ways in which these problems have been tackled. Add any extra ways which *you* have thought of to these contained in this unit.
  a traffic congestion
  b parking problems
  c danger and inconvenience
  d environmental problems.

5 Discover for yourself what parking restrictions are imposed by the different yellow kerb-side markings.

6 Suggest reasons why:
  a Shop-keepers were very worried by the idea of pedestrian precincts.
  b Shoppers are generally in favour of pedestrian precincts.
  c Many shop-keepers now find that pedestrian precincts have *increased* their trade.

7 The following questions are based on Fig. 7.3:
  a Name two squares around which a one-way traffic flow system operates, then state whether both flows are clockwise or anti-clockwise.
  b Name the road which has a large underground car park on its western side, then name the large feature which made it fairly easy to site this type of car park there.
  c Name one road closed to most forms of through-traffic for 12 hours every day, except Sundays. Which two forms of traffic are *excluded* from this restriction? Why are they likely to have been excluded?
  d State what measures have been taken to ease the flow of traffic around Marble Arch.
  e Name one large hotel on Park Lane. Suggest reasons why many high-class hotels are on this road, in spite of the heavy flow of traffic along it.
  f You wish to drive from the US Embassy in Grosvenor Square to the Londoner Hotel on Welbeck Street. Use the scale given in the caption to work out the shortest routes between these two buildings:

  ☐ by following the traffic restrictions shown on the map.
  ☐ by assuming that there are no such restrictions.

  g Why are one-way systems so common, although they force drivers to make longer journeys?

8 Produce a wall display for your classroom based on the theme *How road traffic affects our lives* Try to make your display as useful as possible by including sketches, cartoons and photographs. You could:

  ☐ label a street map of your local town centre to show one-way traffic flows;
  ☐ draw a plan of a local pedestrian precinct;
  ☐ draw graphs based on a questionnaire put to shoppers and shop-keepers in a pedestrian precinct to obtain their reactions to this type of development.

7.2 Rapid transit at home and abroad

In large conurbations, traffic congestion has become so acute that some commuters prefer to travel by rail. People have tended to live further away from the city centres and so the proportion of commuters travelling *long* distances to work has steadily increased. Railways have provided a valuable service for many of these people.

Railways have also played an important role in transporting people *within* city areas, and long sections of them are routed underground in the central districts. London's Underground was one of the first of many such urban railway networks, and provided the model for similar systems throughout the world. They are fast, efficient, produce little pollution and can carry large numbers of people without disrupting other land uses. This unit describes two examples of the **rapid transit system** type of transport network – one in North-East England, the other on the west coast of the United States.

The Tyneside Metro

Tyneside is the most densely populated region in the North-East. At its centre is the dynamic city of Newcastle, an ancient settlement which dates from Norman times. Newcastle's growth was most spectacular during the nineteenth century, when the steel-making, ship-building, engineering and chemical industries drew people into the town in search of work. A narrow gorge made it difficult to build factories in the central area, so most of them were located further upstream or by the wider estuary near the sea. Many of these factories have now closed down, but the areas of housing built around them remain. Their inhabitants have had to seek work further from home and so are faced with much longer journeys than before.

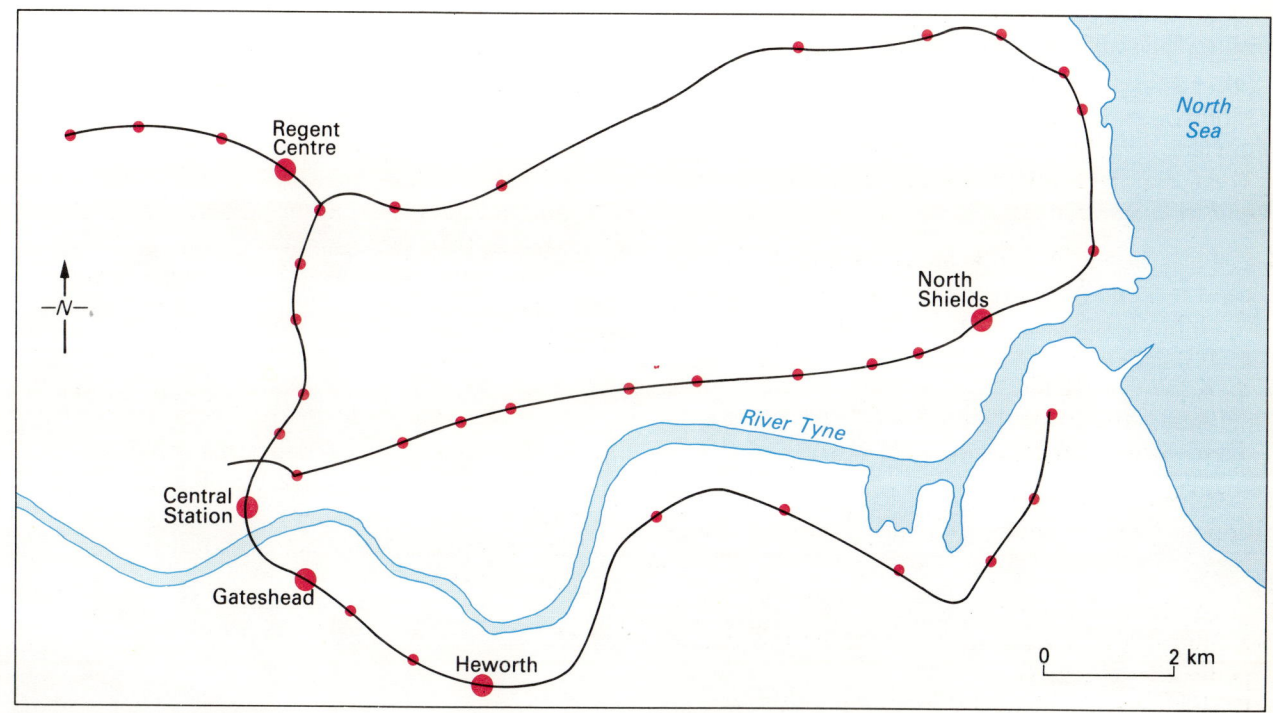

▲ **Figure 7.5** Layout of the Tyneside Metro

Key
— Track
• Station
● Station with interchange to British Rail services

In the 1960s and 70s, the centre of Newcastle was transformed by a massive programme of comprehensive redevelopment. Hundreds of the older houses were demolished to make way for new roads, office blocks, shops and recreational centres. Families who lost their homes were asked to move into new suburbs, some of them a long way from the city centre. They too face longer journeys to work.

Tyneside responded by building a new *internal* railway network. It was at first called the Rapid Transit System, but its name was later changed to 'The Metro'. It was built with the help of a large grant of money from the government, and was opened to the public in 1980.

The Metro has a total length of 55 km (Fig. 7.5). Most of it is on the surface, but the inner city sections have been routed underground. The network's layout was designed to link the industrial zones, the outer suburbs and the new developments in the heart of Newcastle. While some stretches of the route follow disused tracks once owned by British Rail, a number of completely new stations had to be built. The entire network is electrified and specially-designed rolling stock provides a comfortable and highly efficient service for the local people (Fig. 7.6).

The San Fransisco Bay Rapid Transit System

San Fransisco is also at the heart of a large industrial conurbation. It was founded in the eighteenth century by a Spanish mission. It remained quite small until 1848, when gold was discovered to the east of the town. The news travelled very quickly and attracted hundreds of thousands of immigrants from overseas, as well as from the east coast of North America itself. San Fransisco acted as the port for the 'gold rush' which followed, and expanded rapidly along the shores of the great natural harbour named after it.

The feverish activity lasted only as long as gold could be found. Luckily, San Fransisco is also sited near to a great, flat valley which is ideal for growing fruit and vegetables. Many of the immigrants turned to farming; others worked in the docks built to export surplus food. Ship-building and oil-refining are other major occupations in the bay area.

Ideal building land is in short supply. San Fransisco Bay is surrounded by mountain ranges and steep hillsides extend right down to the shore in many places (Fig. 7.7). This has led to high-rise building in the city centre. Further out, vast suburbs of lower-level housing have sprawled to form an almost continuous ring

around the bay. This type of development constantly lengthens journeys to work; it also prompted the region to build a new railway to link all parts of the conurbation (Fig. 7.8).

The first stage of the Bay Rapid Transit System was completed in 1971. It proved so successful that it has since been extended. The aim was to provide commuters and shoppers with a fast and efficient service. Its stations have been widely spaced apart to keep the number of time-consuming stops to a minimum. San Fransisco's system has proved very popular at a time when when most of America's *long-distance* railway routes are losing money.

1 What *is* a 'rapid transit system'?

2 Explain briefly why the Newcastle and San Fransisco areas have attracted such large populations.

3 Suggest reasons why:
a Rapid transit systems are often routed underground in central urban areas.
b The Tyneside Metro generally follows an E-shaped route.
c The San Fransisco Bay Rapid Transit System follows a more circular route.

4 What are the total lengths of:
a Tyneside's Metro (see text)?
b The San Francisco Bay system (see Fig. 7.8)?

5 Suggest what benefits and problems rapid transit systems are likely to produce for:
a City/regional transport authorities.
b People living near to them (some, but not all of them will be commuters).

◀ **Figure 7.6** New rolling stock on Tyneside's Metro

▲ **Figure 7.7** Steep hillsides making building difficult on the shores of San Francisco Bay

▶ **Figure 7.8** The San Francisco Bay Rapid Transit System network

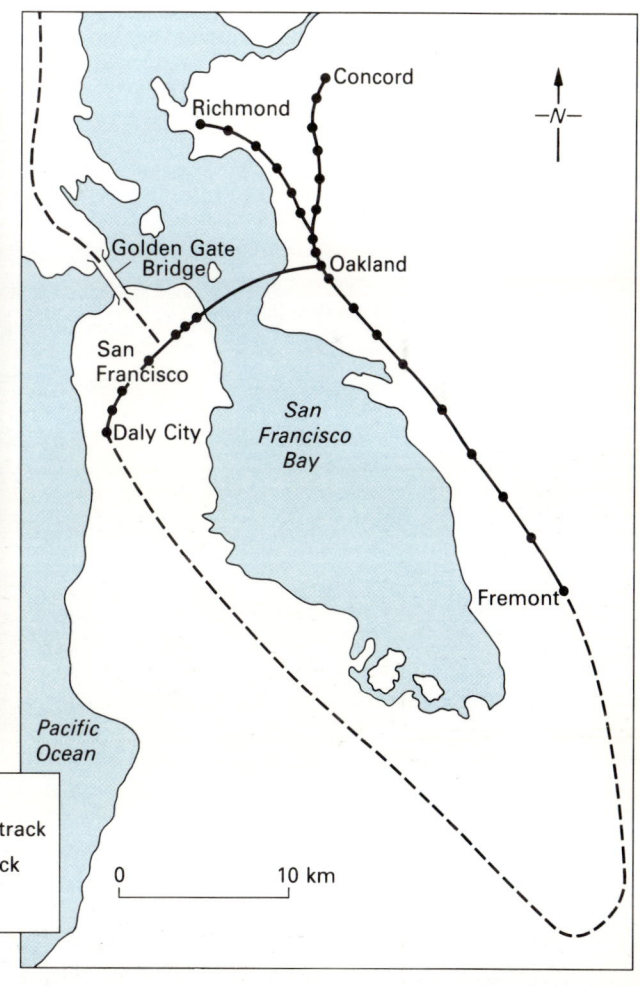

Key

——— Completed section of track

– – – Planned section of track

● Station

0   10 km

7.3 By-passing the problem

By-passes are a very effective way of keeping through-traffic clear of town centres. Examples can be seen throughout the country, and 35 new by-passes were approved by the Department of Transport in 1986 alone. Ring roads are by-passes which encircle a settlement, and the best-known British example is the M25 motorway around London (Fig. 7.9).

The rest of this unit examines the proposal to build a by-pass around an *imaginary* town (so don't try to find it in your atlas!). It looks at many of the issues which planners need to consider when discussing the route for a new by-pass; it will also help you to appreciate how this type of road development can change a community's way of life.

Study Fig. 7.10 very carefully. It shows the position of 'Congeston', the town to be by-passed, and two routes from which one has to be chosen. Congeston is a small market town serving the farming area around it. Its old narrow streets can no longer cope with the volume of traffic using the A1 main road. The B1111 is not quite so busy, but it too becomes very congested during rush hour peak travelling times.

1 a What is the main reason for building by-passes and ring-roads?
b What is the *difference* between a by-pass and a ring-road?

2 Use Fig. 7.9 to answer these questions
a What is the *direct* distance (through the centre of London) between Potters Bar and Redhill?
b How long would this direct journey take at an average speed of 40 kph?
c What is the distance between the same two places *along* the *eastern* section of the M25 motorway?
d How long would this motorway journey take at an average speed of 120 kph?
e What are the main advantages and disadvantages of choosing the M25 route to travel between Potters Bay and Redhill?

The next two questions are based on Fig. 7.10.

3 a Copy out Fig. 7.11. Complete it by putting ticks in the last two columns (to show the *advantages* of each proposed by-pass route).
b According to your completed table, which proposed route has more advantages?
c List the *five* considerations listed in the table which you feel are *most* important. Try to add reasons to justify your selection.

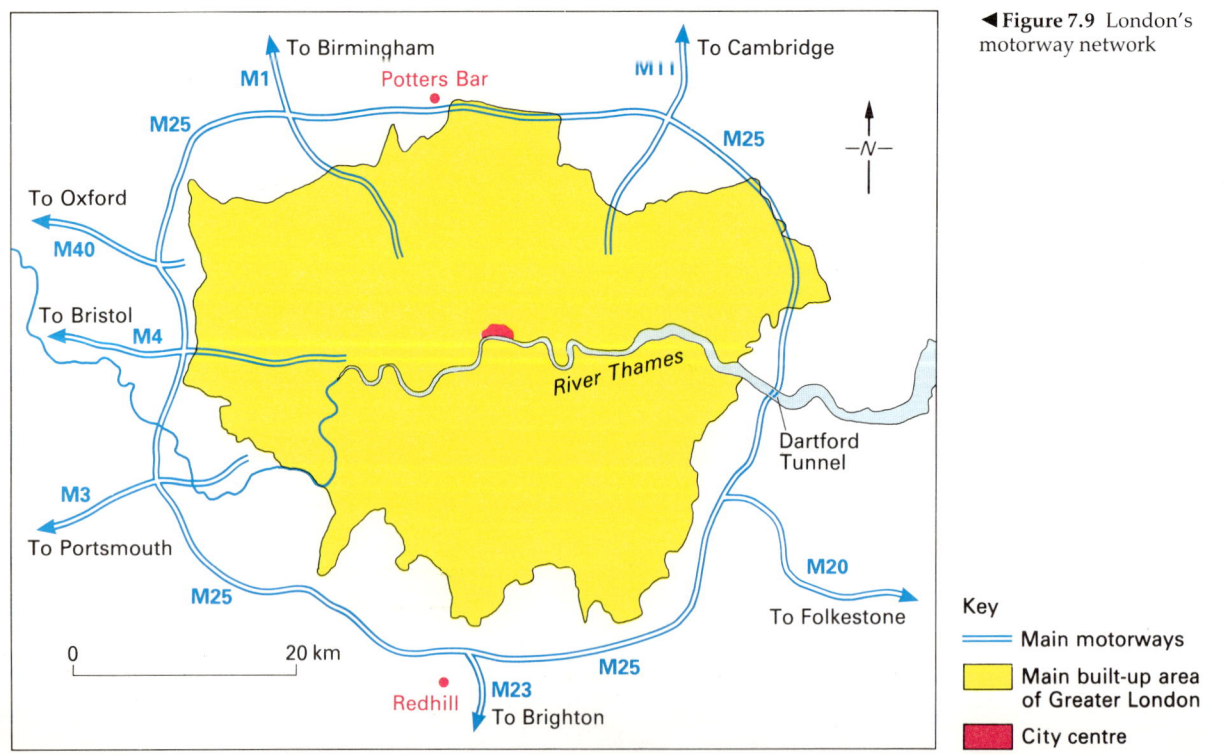

◀ **Figure 7.9** London's motorway network

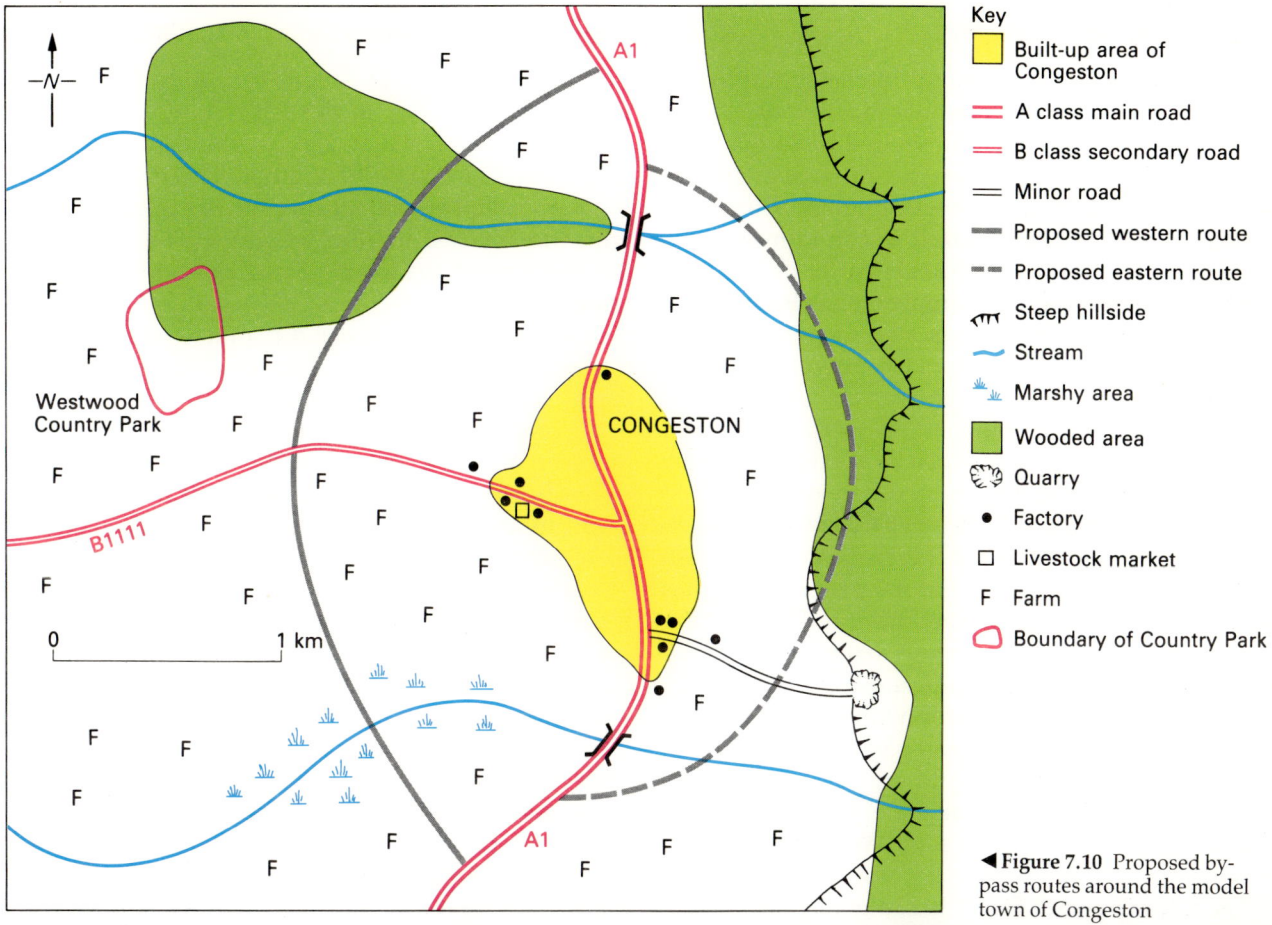

◀ **Figure 7.10** Proposed by-pass routes around the model town of Congeston

▼ **Figure 7.11**

Considerations	Western route	Eastern route
Is the shorter route		
Avoids steep slopes		
Avoids marshy areas		
Avoids the best farmland		
Crosses fewest streams		
Links the most main roads		
Requires fewer trees to be cut down		
Best serves the quarry		
Best serves the country park		
Best serves the livestock market		
Best serves the industrial areas		

4 After class discussion, suggest the most likely reactions of the following to *any* suggestion to build a by-pass around Congeston:

a The manager of the wildlife reserve.
b The owners of farms near to the town.
c The residents of Congeston.
d The Head Teacher of Congeston High School.
e The owner of a garage on the B1111.
f The owner of a public house in Congeston.
g The manager of the country park.
h The local members of the Council for the Protection of Rural England.
i The manager of a road haulage company.

5 a Draw a sketch map of part of your local area which shows *either* the route of a by-pass which has already been built *or* the route you would follow to build a new by-pass around a town or small village.
b Briefly describe the by-pass route shown by your sketch.
c Suggest the main advantages of this by-pass route.
d Does it appear to have any *dis*advantages? If so, what are they?

7.4 Network analysis

Britain's road network is constantly being improved. This is a costly process, not only in terms of money, but also the amount of land required to do this. Our motorway network alone has devoured about 25 000 hectares, mainly in fertile lowland areas. This is the equivalent of 400 average-sized British farms!

Our demand for better roads is partly due to a remarkable increase in the number of vehicles since the Second World War. It is also to provide more direct routes between our main towns and cities. This unit examines ways in which we can find out how efficient our present transport networks really are. The methods described in it can be applied to railway and airline routes as well as road networks.

Study Figs 7.12 and 7.13, which show five imaginary towns and the roads which link them. Both maps are of the same area, but the road network has been shown differently in each case. In Fig. 7.12 the actual routes taken by the roads have been shown. Figure 7.13 is called a **topological map**, which means that all the roads on it are drawn as straight lines. These roads cannot be drawn accurately to scale and so it has been necessary to add the distance between each pair of towns. The numbers in brackets give the populations of these towns (e.g. Westham has 5 000 people).

The information in Fig. 7.13 can tell us how accessible each town is, but first it has to be recorded in the form of a **matrix** (a special kind of table). Two types of accessibility matrix are used in this unit.

The shortest-path matrix

The matrix in Fig. 7.14 is quite easy to complete. The number in each box of the table shows how many stretches of road *between pairs of towns* have to be used to get from one town to another. These stretches of road are called **edges**, and we always take the route with the smallest number of edges. The dots show the positions of the towns. These positions are called **nodes**.

Business people and town planners often need to know how accessible a place is. This means how easily it can be reached by travellers from other places in the area around it. An accessible town is one which has been built in a convenient position, and so is easy to reach; inaccessible towns are much more difficult to get to and so companies will not be very keen to build new offices and factories in them.

It is easy to complete the line for Bigton, as all the other towns in the network can be reached by travelling on just one stretch of road. Bigton-Bigton has to be given a 0 because no travelling is involved. Most of the others in the matrix have a 1 or 0 for the same reasons. There is a choice of two routes in going from Eastby to Northley (via Bigton or Hilling). Both routes have two edges, so Bigton-Northley has been given a 2. There are a number of ways of getting from Eastby to Westham. The 'shortest-path' (most direct) route is via Bigton, which has only two edges. Now check the entries along the matrix lines for Hilling, Northley and Westham, and make sure you understand how all their entries have been obtained.

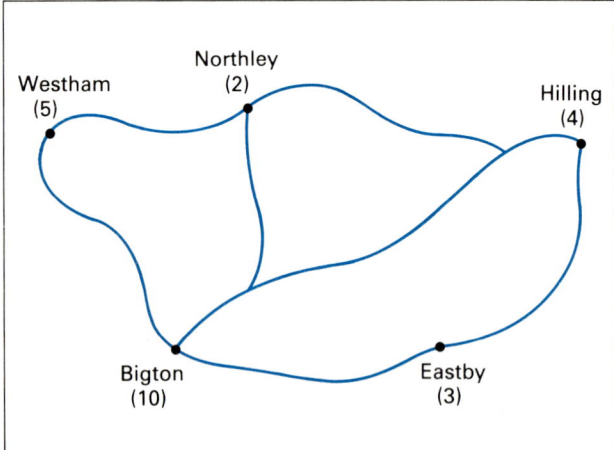

▲**Figure 7.12** A typical route network

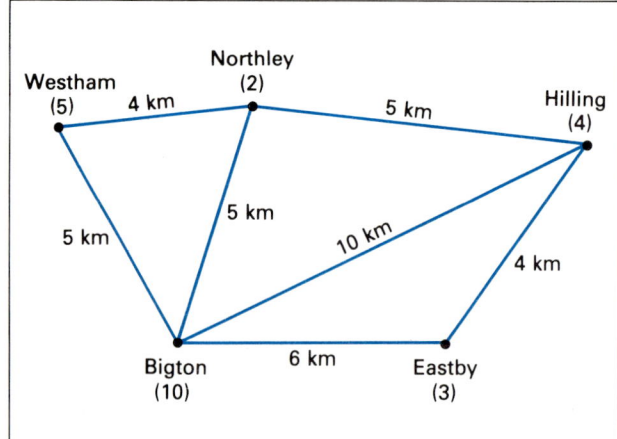

▲**Figure 7.13** Topological map of the network shown in Fig. 7.12

To → From ↓	Bigton	Eastby	Hilling	Northley	Westham	Total (Shimbel number)
Bigton	0	1	1	1	1	4
Eastby	1	0	1	2	2	6
Hilling	1	1	0	1	2	5
Northley	1	2	1	0	1	5
Westham	1	2	2	1	0	6

◀ **Figure 7.14** Shortest-path matrix

The last column in the matrix gives the total number of edges entered along each line. These totals are called **Shimbel numbers**. The town with the smallest Shimbel number (Bigton) is the most accessible, while Eastby and Westham are the most inaccessible places because they both have the largest number. The positions of these towns in Fig. 7.13 suggest that these results are correct. Bigton *must* be very accessible because roads lead from it to all the other towns. Both Eastby and Westham cannot be reached directly from two other towns.

The distance-population matrix

The distance-population matrix is much more useful than the shortest-path matrix because it is based on more detailed information. The entries are obtained by multiplying the road distances between the pairs of places by the populations of the towns being travelled *to*. For example, Bigton-Eastby has 18 because the distance between the two towns is 6 km and *Eastby* has a population of 3000 people.

The last column in this type of matrix follows the usual pattern. Bigton is still the most accessible town in the network because it has the largest population and is quite conveniently located. Put another way, Bigton is the town which can be reached most easily by all the people living in the five towns. Northley is the second most accessible town and so business people are highly likely to build new factories and offices either in Bigton itself, or somewhere close to it along the road leading to Northley. Hilling is the most *in*accessible town. It has the highest total in the last column and is some distance away from the two largest towns.

Network efficiency

The rest of this unit considers the *routes*, rather than the towns which they link together. The **Beta Index** formula shown below provides a simple way of measuring the extent to which people can move *directly* from one town to another. The ideal situation is when each town has its own direct link with every other town, but many factors such as the relief of the land and the high cost of route construction often make this impossible.

$$\text{The Beta Index} = \frac{\text{Number of edges in the network}}{\text{Number of nodes in the network}}$$

To → From ↓	Bigton	Eastby	Hilling	Northley	Westham	Total
Bigton	0	18	40	10	25	93
Eastby	60	0	16	18	55	149
Hilling	100	12	0	10	45	167
Northley	50	27	20	0	20	117
Westham	50	33	36	8	0	127

◀ **Figure 7.15** Distance-population matrix

Worked example:

In Fig. 7.14 there are five towns (nodes) and seven edges (routes) linking them. The Beta Index for this network is therefore 7/5 = 1.4. This is not a very high value, showing that the network can be improved. To achieve the ideal, three more road links would have to be built. They are Eastby-Northley, Eastby-Westham and Hilling-Westham. The Beta Index for this 'perfect' network is 10/5 = 2.0.

Finally, we can use these two Beta Index calculations to work out the *efficiency* of the network without these extra roads. The formula needed to do this is:

$$\frac{\text{network}}{\text{efficiency}} = \frac{\text{Beta Index for the present network}}{\text{Beta Index for the 'perfect' network}} \times 100$$

This formula gives a percentage answer. In this worked example it is (1.4/2.0) × 100 = 70%.

It is also possible to assess the efficiency of *individual routes*. This is done by comparing the actual distance along a route with its straight-line direct distance, using this formula:

$$\frac{\text{route}}{\text{efficiency}} = \frac{\text{straight-line distance}}{\text{actual distance}} \times 100$$

This formula also gives a percentage answer. For example, we know that the actual distance between Bigton and Hilling is 10 km. If the straight-line distance between them is only 8 km, the route efficiency must be (8/10) × 100 = 80%.

This kind of information is very useful to route planners. It tells them which stretches of road most need to be straightened or replaced because they are so costly and time-consuming to use.

1 Explain very carefully what these terms mean:
   a accessibility
   b accessible
   c inaccessible
   d a route network
   e a topological map
   f a matrix
   g the Shimbel number
   h edges
   i nodes
   j the Beta Index
   k network efficiency.

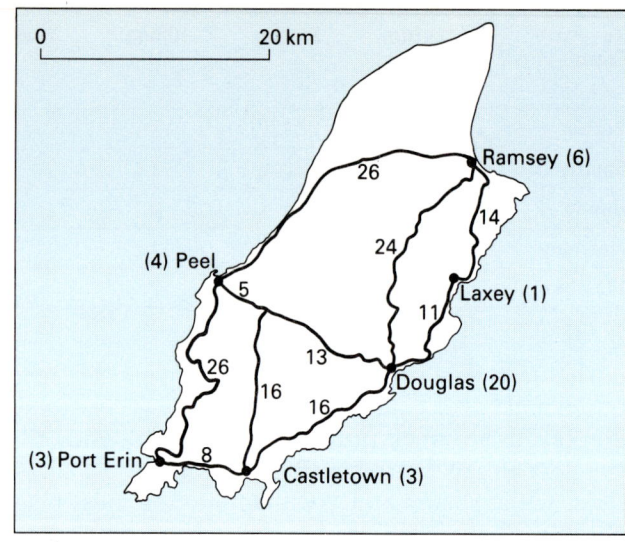

▲ Figure 7.16

2 **a** Trace this map of the Isle of Man's main roads and six largest towns, but *re-arrange the road network into a topological layout*. The 21 km long Peel-Castletown route should also be drawn as a straight line, even though part of it is shared with the road between Peel and Douglas.
   b Draw and complete a shortest-path matrix for this network.
   c Name the most accessible and most inaccessible towns in the Isle of Man, according to your matrix.
   d Draw and complete a distance-population matrix for the same network.
   e Name the most accessible and inaccessible towns according to this second matrix.
   f Comment on any differences in your answers to c and e.

3 Calculate the following information for the route network of the Isle of Man:
   a The Beta Index for its present network.
   b The Beta Index for the 'perfect' network.
   c Its network efficiency, using your answers to a and b.

4 **a** Work out the route efficiency percentages for the most direct stretches of road between every pair of towns on the Isle of Man. (Hint: The linear scale will help you to obtain the direct distances you need to do this.)
   b According to your results for a, which of these routes is the most *in*efficient one on the island?

Counter-Urbanisation

8.1 Introduction: outward migration

The migration of people from rural areas has hastened the growth of the world's largest cities. In the developed countries, this migration has taken place over long periods (often up to 200 years). The result has been overcrowding in the older, inner city areas, and many of the families there now wish to move away from them! This outward movement of people is called **counter-urbanisation**. London provides one of Britain's best examples of this, but similar outward migrations occur throughout the developed world.

Figure 8.1 shows that the combined populations of London's inner and outer areas rose until the early 1950s, since when they have declined. The earlier increase was due mainly to inward migration, and the decline which followed it the result of outward migration. In fact, the population of every one of London's 33 boroughs declined during 1971–81.

This outward migration has been strongest from the inner boroughs, e.g. Tower Hamlets – just to the east of the city centre. 597 000 people lived there in 1901. By 1981 this number had dropped to only 143 000, less than one-quarter of the earlier figure.

The *City of London* – the capital's financial centre – provides an even more stunning example of outward migration, for its population in 1961 was only 4% of that in 1801 (Fig. 8.2). This borough's population has however remained fairly steady since 1961, due to some major redevelopments such as the Barbican Scheme which have included some new housing. Every working day, however, over half a million commuters enter the square mile of the City to work there (Fig. 8.3). The headquarters of most of our banks, insurance companies and shipping lines are located in the City and they have created many highly paid jobs. These workers can afford the cost of long-distance commuting from rural areas in South-East England.

Census year	Population of the City of London, to the nearest 1 000
1801	128 000
1841	124 000
1881	51 000
1921	14 000
1961	5 000

▲**Figure 8.2** Decline of the City of London's population between 1801 and 1981

▼**Figure 8.3** Commuters at Liverpool Street station in London's morning rush hour

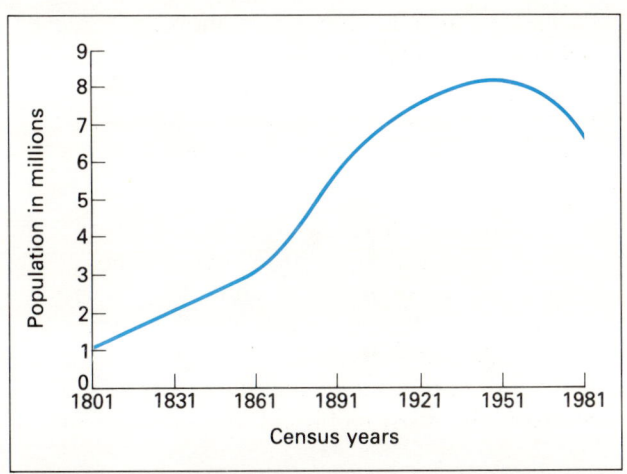

▲**Figure 8.1** Line graph showing changes in the population of Greater London, 1801–1981

Recent improvements in both public and private transport have made the central areas much more accessible (easier to get to). This has increased the value of the land so much that few people can afford to live in them (Fig. 8.4). Government departments, local councils and large companies are able to pay the high rents; they are also quick to take control of land on which the oldest housing stock has deteriorated beyond repair and therefore been demolished. Postwar road improvements and car parking schemes have also taken much land within the inner boroughs previously occupied by houses (Fig. 8.5).

Central London is frequently overcrowded. Trains, buses and pavements are packed by throngs of workers and shoppers at the rush hour peak travelling times, which makes movement tedious and uncomfortable. The hectic pace of inner city life does not suit everyone and many people would rather live in quieter places in the country and on the coast.

Traffic noise and pollution tend to be much higher in city centres. London's western inner areas are lucky to have some very large parks; many inner areas do not, and children living there have more opportunity to escape from the traffic and maze of buildings which are so much a part of the London scene.

1 Use Fig. 8.1 to complete this table:

Year	Population of Greater London (to the nearest 100 000)
1801	
1951	
1981	

2 **a** Plot the information in Fig. 8.2 as a line graph.
  b Describe the population changes shown by your graph. You should include some population figures in your description.

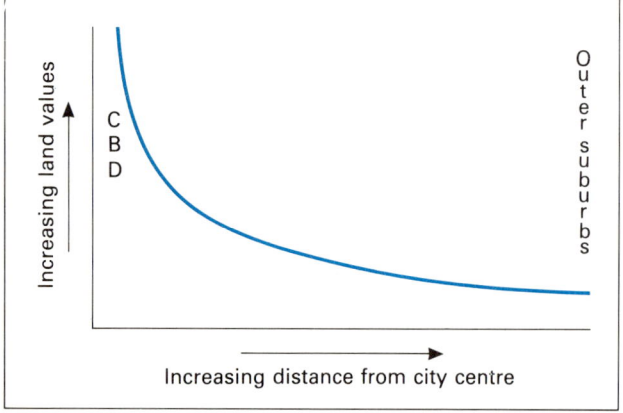

▲ **Figure 8.4** High land values have encouraged many businesses to leave CBDs in favour of cheaper edge-of-town sites

▶ **Figure 8.5** The overhead section of the M4 motorway in the Chiswick area of London – just to the west of the CBD

3 Explain why there is such a great difference between the day-time and night-time populations of the City of London.

4 List as many reasons as you can for the outward migration of people from London's inner boroughs.

5 **a** Shade a map of London's 32 residential boroughs to display the information given in the table on this page. (The City of London has been left out because so few people live there.) The recommended key, colour scheme and title are given below.

(in red)	☐ Boroughs with high population decreases (−30% to −12%).
(in yellow)	☐ Boroughs with moderate population decreases (−12% to −5%).
(leave blank)	☐ Boroughs with low population decreases (−5% to 0.0%).

Title: Population decrease in London boroughs, 1971–81

b Describe the pattern of population decrease shown by your completed map.

c Add a fourth entry to your key:

☐ London boroughs with the ten highest population densities in 1971

d Dot-shade the following ten boroughs on your map:

Camden
City of Westminster
Hackney
Hammersmith and Fulham
Islington
Kensington and Chelsea
Lambeth
Southwark
Tower Hamlets
Wandsworth

e Copy out only the following statements which are *true*, according to your map:

- ☐ All the boroughs next to the City of London are very densely populated.
- ☐ All the boroughs next to the City of London have a high population decrease.
- ☐ The boroughs with the highest population densities have experienced a high rate of population decrease.
- ☐ Two-thirds of all the boroughs with a high rate of population decrease lie by the River Thames.
- ☐ All the boroughs with a low rate of population decrease lie on the edge of the outer boundary of London's built-up area.
- ☐ None of the boroughs with a high density of population have a low rate of population decrease.
- ☐ The higher the population density, the more likely an inner city borough is to shed population through outward migration.

Name of London borough	% Population Change 1971–81
Barking and Dagenham	− 6.6
Barnet	− 4.6
Bexley	− 1.0
Brent	−10.5
Bromley	− 3.6
Camden	−17.0
City of Westminster	−20.5
Croydon	− 5.2
Ealing	− 7.0
Enfield	− 3.4
Greenwich	− 2.7
Hackney	−18.2
Hammersmith and Fulham	−20.9
Haringay	−15.4
Harrow	− 3.4
Havering	− 2.9
Hillingdon	− 2.5
Hounslow	− 3.5
Islington	−20.9
Kensington and Chelsea	−26.3
Kingston upon Thames	− 5.8
Lambeth	−20.1
Lewisham	−13.1
Merton	− 7.0
Newham	−11.8
Redbridge	− 6.2
Richmond upon Thames	− 9.6
Southwark	−19.2
Sutton	− 0.6
Tower Hamlets	−13.8
Waltham Forest	− 8.4
Wandsworth	−15.4

8.2 The rural–urban fringe

Town-dwellers are now making greater use of the countryside than at any time in the past. This unit examines the effects of this increased activity on our rural areas, particularly **rural-urban fringes** on the outskirts of built-up areas. Geographers sometimes describe the fringe areas as being 'rurban'. This means that while they still have some country features such as fields, they are becoming increasingly urban in character – in fact mere extensions of the towns next to them (Fig. 8.6).

There are many reasons for this trend. Town-dwellers now have much more leisure time; many of them also have cars and enjoy driving through nearby stretches of open countryside. Villages within the fringe areas are constantly expanding to house more commuters. These people welcome the chance to 'unwind' away from the hurly-burly of city life, but the demand for a house in the country is now so great that the beauty and quietness which these newcomers seek are being seriously threatened. More commuters means heavier traffic and additional land has to be set aside for road improvements. Many settlements are being by-passed by new ring-roads, which increase noise and pollution levels. They also tend to disrupt farming activities. As fields become 'ripe' for development, farmers reduce their investment in buildings and land. This neglect leads to a rapid deterioration (worsening) of the rural landscape, as farmers patiently await the most profitable time to sell.

Both commuters and town-dwellers place many conflicting demands on rural-urban finges. Schools, hospitals and leisure facilities benefit greatly from the cleaner country air. Recreation is big business today and golf is just one of many such activities which compete for the limited amount of land which is available (Fig. 8.7).

▼ **Figure 8.6** A 'rurban' scene near Kendal

◀ **Figure 8.7** One example of conflict in a rurban area!

1 List separately the rural and urban features within the 'rurban' scene in Fig. 8.6.

2 Describe in some detail the many ways in which our rural-urban fringes are becoming more urbanised. (Try to add some of your own ideas to the information given in this unit.)

3 The extract from a *Daily Telegraph* article (right) is based on a rural-urban fringe area in Kent. It is clear that the writer of the article is worried by the businessman's decision to sell his land in such small units; he is also against this farmland being used for 'horseyculture'.

  a Explain what the term 'horseyculture' means.

  b Imagine you are the businessman described in the extract. You believe that Peter Birkett's article is very biased (one-sided) so you decide to write a letter to the editor of the newspaper giving *your* side of the argument. Your letter could mention the many benefits which horse-riding can bring to people of all ages; it could also state that the alternative is for you to sell your land to house-builders or industrial developers.

  c Now that you have examined the arguments both for and against the proposed sale:

  □ say whether you feel the proposed sale *should* go ahead;

  □ summarise the *main* arguments which led you to this decision.

4 Study the area on the Ordnance Survey map extract on page 148 which is to the *east* of grid line 70. Write about 300 words to assess the accuracy of the statement: 'That this area is more urban than rural in character'.

Culture that threatens the heart of countryside

By Peter Birkett

Thistle-Blown fields littered with shacks, 'day-glo' road cones, baulks of timber and barrels to make jumps for poorman's equestrianism. It is known to the planners, the few that care, as 'horseyculture' and surprisingly most of it seems to be in breach of the law.

Green Belt farmers, frustrated by not being able to cash in on the value of their land for houses, have long encouraged horseyculture by selling an acre or two at premium prices to absentee horse-owners. Now with the price of agricultural land falling and with pony interest at an all-time high, horseyculture has spread from the suburbs to threaten the very heart of the countryside.

A few years ago, a businessman bought 36 acres of land, a beautiful hillside above a wild river valley near my home in Kent. He paid £700 an acre and improved it with an EEC grant.

Now he wants to sell off the land as 11 'fenced and serviced' grazing plots for horses at more than £2 500 an acre, which is £1 000 an acre more than local farmland market prices.

The few people living thereabouts were worried. We feared for the plovers and the partridges nesting on the hillside and for the kingfishers and occasional otters down by the river. We feared too that the land, once broken up into piecemeal ownership, could never be restored to farmland ...... Sadly, horseyculture is causing more damage to our landscape than oil exploration.

8.3 Broughton: a suburbanised village

Broughton, in Lancashire, is now a very popular **dormitory settlement**. This means that many of its inhabitants commute daily to other, larger, towns to work. Figure 8.8 shows that Broughton is situated (located) to the north of Preston and is only a short distance from this important industrial centre. Preston also acts as the 'county town' of Lancashire and so handles much of its routine administration. A narrow stretch of open countryside separates the two settlements, but Broughton is virtually a suburb of Preston.

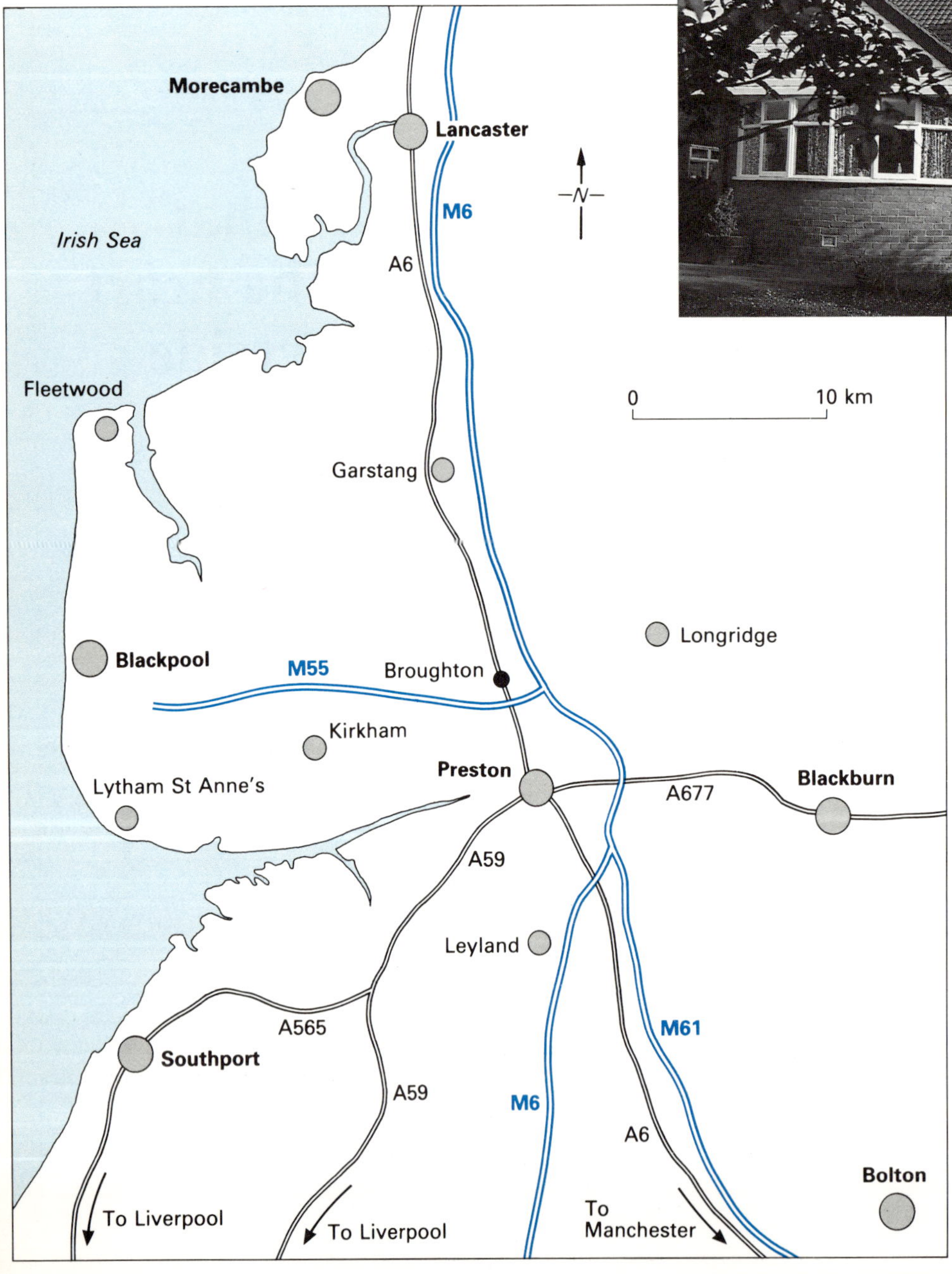

◄**Figure 8.8** The situation of Broughton

▲**Figure 8.9** (top)

▲**Figure 8.10**

Morecambe
Lancaster
M6
A6
Irish Sea
Fleetwood
0   10 km
Garstang
Longridge
Blackpool
M55   Broughton
Kirkham
Lytham St Anne's
Preston
Blackburn
A677
A59
Leyland
A565
M61
Southport
A59
M6
A6
Bolton
To Liverpool
To Liverpool
To Manchester

Key

● Large town centre	**Large town centre**
● Small town centre	Small town centre
● Site of Broughton	Site of Broughton
Motorway	Motorway
A class main road	A class main road

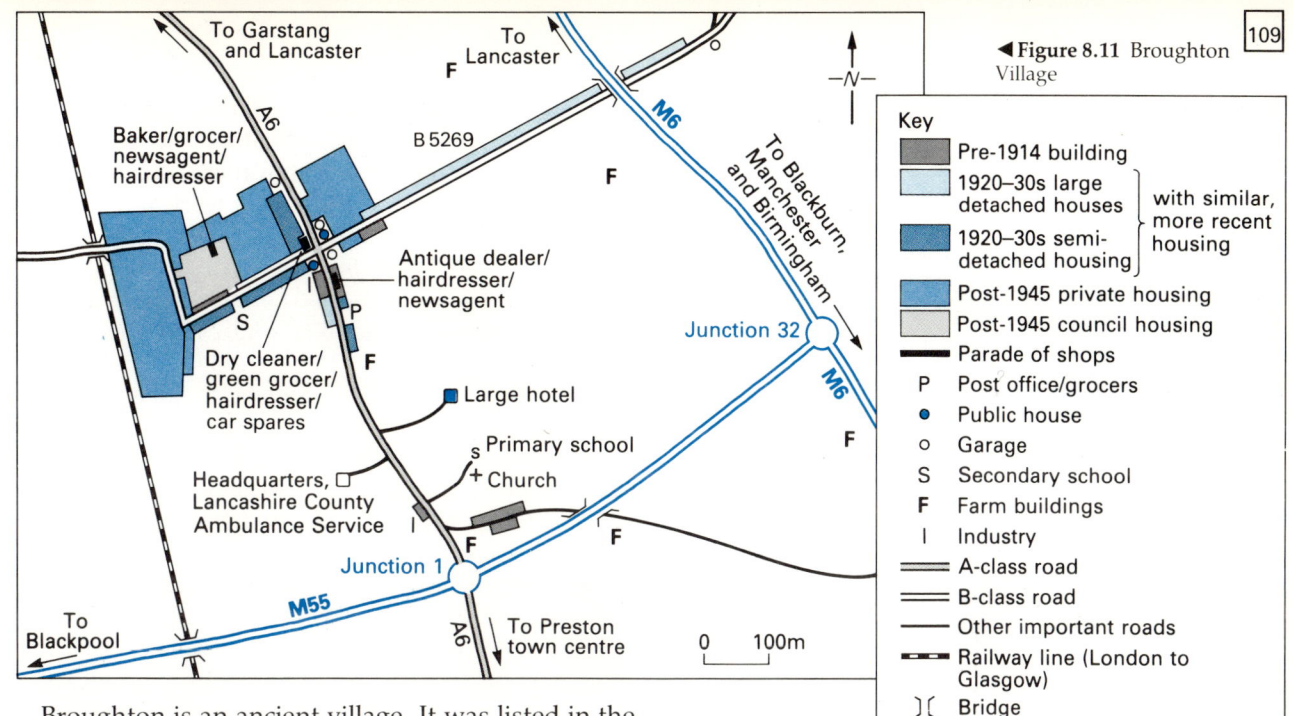

◀ Figure 8.11 Broughton Village

Key

▨	Pre-1914 building
▨	1920–30s large detached houses
▨	1920–30s semi-detached housing

with similar, more recent housing

▨	Post-1945 private housing
▨	Post-1945 council housing
▬	Parade of shops
P	Post office/grocers
●	Public house
○	Garage
S	Secondary school
F	Farm buildings
I	Industry
═══	A-class road
══	B-class road
──	Other important roads
▪▪▪	Railway line (London to Glasgow)
)(	Bridge

Broughton is an ancient village. It was listed in the Domesday Book, compiled by the Normans shortly after they invaded England in AD 1066. Its main role over the last 900 years has been to meet the needs of the surrounding farming area. It remained a small village until the present century. Increasing car ownership then made it possible for wealthy Prestonians to live in the country, yet within easy reach of their offices and factories. This created a demand for expensive detached houses (Fig. 8.9) which were built along Broughton's main roads.

Continued outward migration has led to many semi-detached bungalows being built on small estates just off these roads (Fig. 8.10). The building of the M6 motorway has greatly increased Broughton's popularity and some of its workers now travel to Blackburn and even as far as Manchester.

The map in Fig. 8.11 shows the present layout of Broughton Village. Note that the oldest buildings (the church and the primary school) are some distance from the new housing areas. The new areas are clustered around the crossroads where the B5269 joins the much busier A6 main road.

1 What is meant by the term 'dormitory settlement'?

2 a What evidence shows that Broughton is an ancient settlement? (Hint: documents, buildings.)
b What was the original role of Broughton Village?
c Suggest why Broughton's new housing estates are some distance from the church and the primary school.
d Give at least four reasons why Broughton has become very popular with commuters.

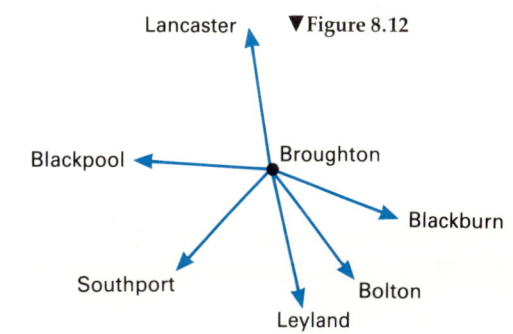

▼ Figure 8.12

3 a Copy Fig. 8.12 which shows the situation of Broughton Village.
b Label each route-line on the diagram with

☐ its compass direction *from* Broughton.
☐ the *direct* distance (in km) from Broughton to the town named on it.

4 Broughton is still only a fairly small village of fewer than 1 000 people. Why, therefore, does it have:
a So many garages? (How many?)
b More than one public house? (How many?)
c A large 'high-class' hotel equipped with a swimming pool and fitness centre?
d A modern secondary school able to take 800 pupils?
e *Four* hairdressing salons (including one not shown on Fig. 8.11)?

5 Suggest reasons why Broughton does *not* have its own supermarket, bookshop or hardware store?

8.4 Green belts and new towns

Our towns and cities sprawled (grew outwards) very rapidly during the 1920s and 1930s and this became a matter of great concern. In 1940 the Government appointed the Barlow Commission to suggest ways in which future urban sprawl might be controlled. The members of the Commission recommended that **green belts** should be established around the major cities.

These areas of countryside would be protected by law, and only limited industrial and residential development be allowed within them. It was hoped that this action would preserve the rural character of the belts, provide 'breathing spaces' between the cities, *and prevent them from joining up* to form huge, continuous built-up areas. London was the first British city to adopt the idea, and established a belt 15–20 km wide around itself in 1944.

It was soon realised that green belts alone could not solve the problem of urban sprawl. The cities still needed to build new houses and these would take up much more space than the old ones which they replaced. The planners feared that the sprawl would simply continue *beyond* the belts and so force commuters to travel even further to their places of work in the cities.

The problem of re-housing city-dwellers was first tackled by the New Towns Act of 1946. It permitted a ring of new towns to be built around the outer edge of the green belt. Each town would accommodate about 50 000 people and be completely self-contained; it would have all the houses, factories and services needed by its inhabitants. This idea was not completely new, however, for Leonardo da Vinci had previously suggested a similar plan in the fifteenth century to cope with serious overcrowding in Milan (in northern Italy).

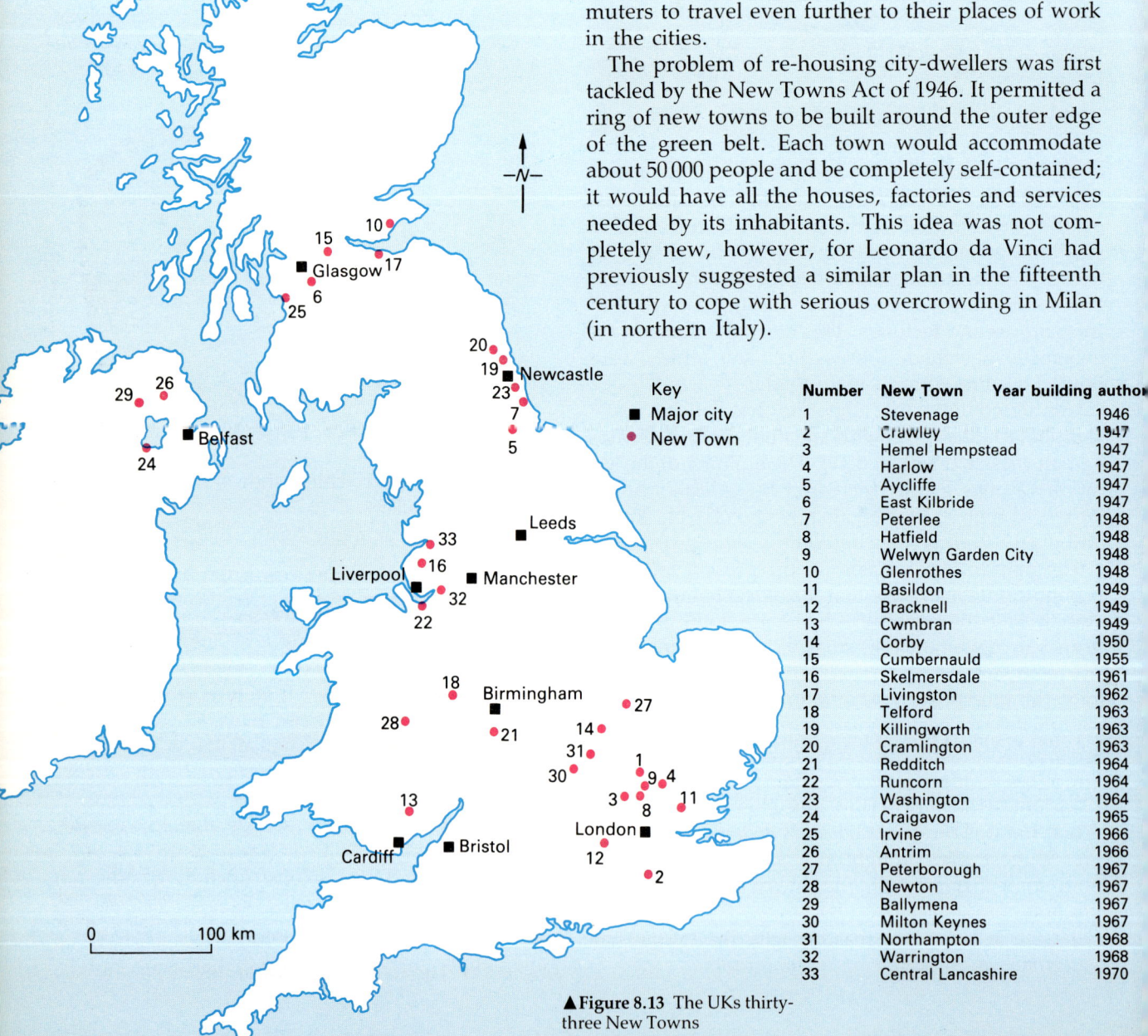

Number	New Town	Year building authoris
1	Stevenage	1946
2	Crawley	1947
3	Hemel Hempstead	1947
4	Harlow	1947
5	Aycliffe	1947
6	East Kilbride	1947
7	Peterlee	1948
8	Hatfield	1948
9	Welwyn Garden City	1948
10	Glenrothes	1948
11	Basildon	1949
12	Bracknell	1949
13	Cwmbran	1949
14	Corby	1950
15	Cumbernauld	1955
16	Skelmersdale	1961
17	Livingston	1962
18	Telford	1963
19	Killingworth	1963
20	Cramlington	1963
21	Redditch	1964
22	Runcorn	1964
23	Washington	1964
24	Craigavon	1965
25	Irvine	1966
26	Antrim	1966
27	Peterborough	1967
28	Newton	1967
29	Ballymena	1967
30	Milton Keynes	1967
31	Northampton	1968
32	Warrington	1968
33	Central Lancashire	1970

▲Figure 8.13 The UKs thirty-three New Towns

▲ **Figure 8.14** 'Old' and 'New' Edinburgh

The Town Development Act of 1952 proposed another way of solving the problem. This Act gave the Government the power to enlarge *existing* towns to re-house people from the great cities. These **expanded towns** were especially attractive to the planners because they already had many basic services and so would be much cheaper to complete. The expanded towns weren't a totally new idea either! Edinburgh had become dangerously over-crowded by the mid-eighteenth century and in 1767 Parliament approved an extension to the north of the old town and the castle. The newer area can be clearly seen in Fig. 8.14. Princes Street, in the centre of the picture, is the best-known of many delightful new roads built at that time.

More than thirty new and expanded towns have been built in the United Kingdom over the last 40 years (Fig. 8.13), and about 5% of all our people now live in them. Most of these towns were intended to take the **overspill population** from the large cities, but some have met other needs. Corby, for example, used to be a quiet village surrounded by pleasant countryside. Then large deposits of iron ore were discovered nearby and a steel works was built there to process it. The village was expanded into a thriv-ing town of some 50 000 people. The steel works has since closed down, but the new town remains. Peterlee and Glenrothes were built to improve the quality of life of miners in old, decaying villages on two of our coalfields.

Plans for the first group of new towns were approved by Parliament between 1945 and 1950. Their names and positions are shown in Fig. 8.13. Most of these earlier towns formed a ring around London's green belt. They all shared the same basic design and were intended to re-house ordinary working people. The areas within them set aside for housing, industry and recreation were usually kept well apart. The houses were grouped together into separate suburbs called **neighbourhood units** (Fig. 8.15). Each unit had 5 000–8 000 people, which is about the same size as a small country town, and was provided with its own primary school, pub, church, 'village hall', and a parade of small shops to meet day-to-day needs. The main public buildings, super-markets and shops selling specialised goods such as furniture were located in the centre of the town. The factories were also grouped together – into industrial estates.

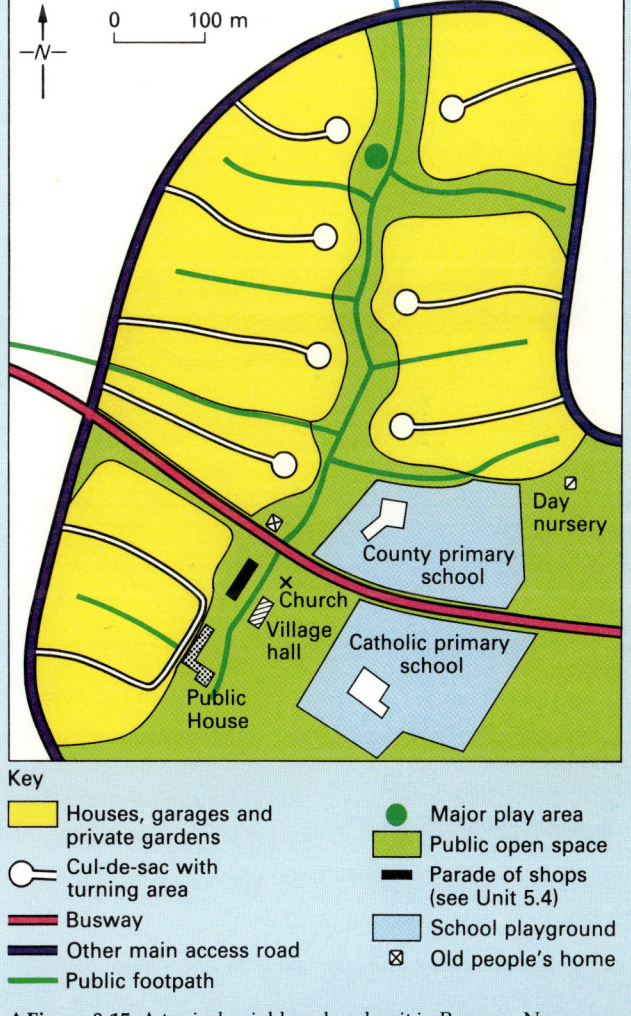

Key

🟨	Houses, garages and private gardens
⊸	Cul-de-sac with turning area
▬	Busway
▬	Other main access road
▬	Public footpath

🟢	Major play area
🟩	Public open space
▬	Parade of shops (see Unit 5.4)
🟦	School playground
⊗	Old people's home

▲ **Figure 8.15** A typical neighbourhood unit in Runcorn New Town

A second and much larger group of towns was approved after 1950. These followed a variety of basic designs, some of them very 'modern' indeed, e.g. Runcorn (also described in Units 5.4 and 8.5). These later towns also provided houses for the 'better-off' and so some of their neighbourhood units were planned to have quite low population densities. Many of the houses in these new towns are now owner-occupied.

The towns in the second group took road safety very seriously. Runcorn, for example, has two completely separate networks of main roads. The Expressway is a two-lane motorway for lorries, coaches and cars which follows a 'figure of eight' pattern. It links Shopping City (see Unit 5.4), the industrial estates and the various housing estates, but does not pass through them. Runcorn's Busway gives more direct access to factories and homes, but can be used only

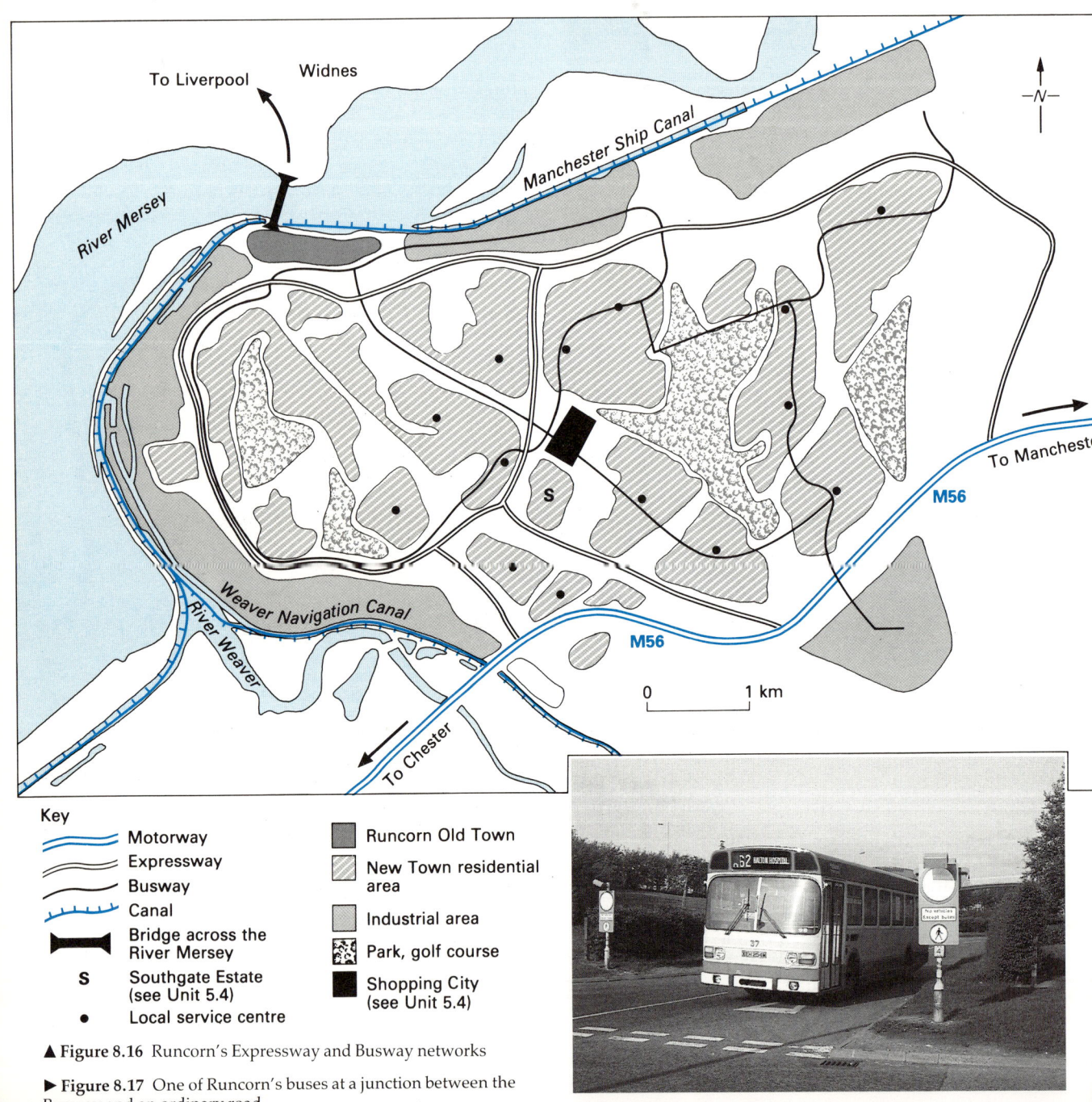

Key

〰️ Motorway

〰️ Expressway

〰️ Busway

⊢⊢⊢⊢ Canal

▬ Bridge across the River Mersey

S Southgate Estate (see Unit 5.4)

● Local service centre

Runcorn Old Town

New Town residential area

Industrial area

Park, golf course

Shopping City (see Unit 5.4)

▲ **Figure 8.16** Runcorn's Expressway and Busway networks

▶ **Figure 8.17** One of Runcorn's buses at a junction between the Busway and an ordinary road

by the council's buses. Traffic lights automatically give them right of way where the Busway crosses minor roads. Runcorn's buses can drive twice as fast as those in our older towns because they don't have to compete with other kinds of traffic. Towns like Runcorn have made use of paths, footbridges and under-passes to segregate (keep apart) pedestrian and vehicle traffic.

The story of these towns has not been a completely happy one. Problems have occurred because people moved into them before all the recreational and other facilities they needed had been completed. There were many social problems too. You can imagine what it must be like to leave your relatives and close friends behind and go to a place where every face is 'new'. It may take you a long time to find trusted baby-sitters making it difficult to meet people socially. It also takes time for new communities to develop their own clubs and societies. The result is boredom, frustration and vandalism. The crime, divorce and suicide rates in these towns have been somewhat higher than for the country as a whole.

Jobs are vital to the success of any new community. The Government has therefore offered special incentives to companies wishing to open factories in the new and expanded towns. These have included:

□ grants towards the cost of new buildings and machinery;
□ grants to reduce the cost of re-training key workers;
□ low factory rents;
□ reduced taxation for the first few years.

Tragically, many factories in these towns closed down shortly after their benefits stopped. The result has been high unemployment, a deep feeling of resentment and continued dependence on the cities where the migrants came from.

The government has invested hundreds of millions of pounds in these towns, but the rate of investment in them is now much lower. This is partly because the United Kingdom's population is growing more slowly; also because riots in some of our inner city areas forced the Government to put more money into these areas in the 1980s.

1 a Explain the meaning of the term 'overspill population'.
b Complete this table, with the help of Fig. 8.13.

Name of large British city	Name of one new town built to take the overspill population from this city
Belfast	
Birmingham	
Cardiff	
Glasgow	
Liverpool	
London	
Newcastle	

2 a After class discussion, explain why many of our new towns have been divided up into 'neighbourhood units'.
b Imagine that you have just moved into a new town. Write an entry for your diary which describes your feelings about the move; try to give a balanced account, which includes both the pleasures and the worries which the move has given you.

3 Write a detailed essay entitled 'British Green Belts and New Towns' which includes information on the following topics.

□ Recommendations of the Barlow Commission of 1940.
□ Purposes of the New Towns Act of 1946 and the Town Development Act of 1952.
□ Reasons why expanded towns are especially popular with the planners.
□ Main characteristics of the first group of British new towns.
□ Ways in which the later new towns are different to those of the first group.
□ Government help for factories established in the new towns.
□ Reasons why less public money is now being spent on our new towns.

4 Describe the layout of Runcorn New Town, under these separate headings:
a road networks
b industrial locations
c residential areas
d recreational areas.

8.5 Migration to Runcorn New Town

Most migrants to Runcorn New Town have come from the Merseyside area, particularly the inner city districts of Liverpool (see Unit 2.6 on Toxteth). However Runcorn has welcomed people from all over the country (Fig. 8.18); it has also tried to attract people from all walks of life – not just those whose houses have been demolished as a result of urban renewal schemes. Its planners wished to create a 'balanced' community which includes people of all ages and in many different types of work. Families who can afford to buy their own homes are just as welcome as those who can't. Runcorn has spent a considerable sum of money on television advertising and publicity leaflets. The following extract is taken from a free leaflet aimed at owner-occupiers – those who wish to buy their own homes.

Origins of immigrants	1982	1984
Liverpool	50%	40%
Rest of Merseyside	16%	23%
Manchester	0%	2%
Lancashire	4%	0%
Cheshire	12%	14%
London	2%	2%
Other areas	16%	19%
Total	**100%**	**100%**

▲**Figure 8.18** Table showing the origins of Runcorn New Town's immigrants

▶**Figure 8.19** 'People can relax in Runcorn New Town'

'You're already well down the path to making one of the best financial decisions of your life, because you're giving serious consideration to buying a new house – and by looking in the Runcorn area, you're getting even more things in your favour.

'Few areas can offer you such a wide choice of properties at prices to suit most pockets – and there are extra advantages in Runcorn which you may not know about.

'Take, for example, its location – close enough to the big cities of Manchester and Liverpool and other major towns to offer rapid access to all their facilities, yet far enough away to enjoy fresh, open countryside. Runcorn itself has a range of shopping, sports and leisure facilities which are the envy of many larger towns.'

This material is attractively illustrated, with many carefully chosen views of relaxed people enjoying a 'typical' summer's day in Runcorn! (Fig. 8.19). The leaflet then lists five quite different types of house which can be *bought* in Runcorn:

☐ one-bedroom flats
☐ two-bedroom bungalows
☐ compact town-houses
☐ two and three-bedroom houses
☐ luxury four-bedroom detached houses (Fig. 8.20).

▼Figure 8.20

▲ Figure 8.22 Southgate flats, Runcorn New Town

Figure 8.21 compares house tenure in Runcorn New Town with that for the whole of Cheshire. Southgate has the highest percentage of rented accommodation in Runcorn; it also has the highest population density, although there are some spacious public gardens and play areas between the blocks of flats (Fig. 8.22). The units in these blocks range from one-bedroom/two-person flats to much larger ones able to sleep five or six people (see Fig. 8.23). Originally all Southgate's flats were rented out by the local council; a few of them have now been sold to residents wishing to become owner-occupiers.

▼Figure 8.23 Five person, three bedroomed maisonette

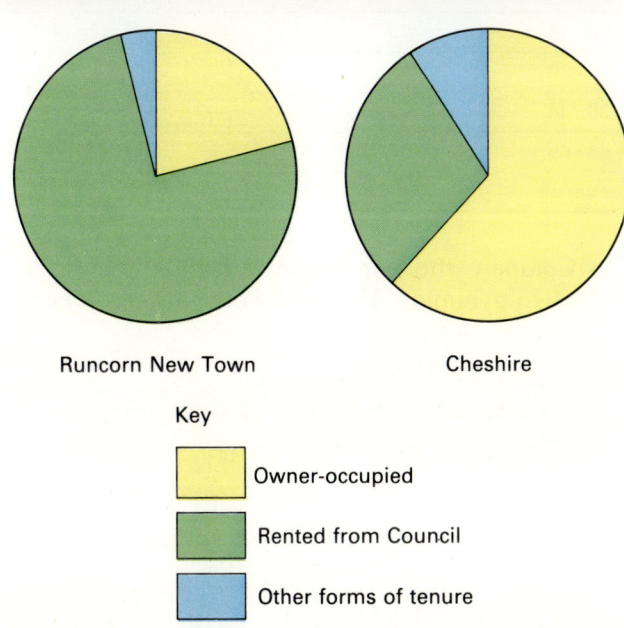

Runcorn New Town   Cheshire

Key

■ (yellow) Owner-occupied

■ (green) Rented from Council

■ (blue) Other forms of tenure

▲Figure 8.21 Pie graphs showing house tenure in Runcorn New Town and the country of Cheshire

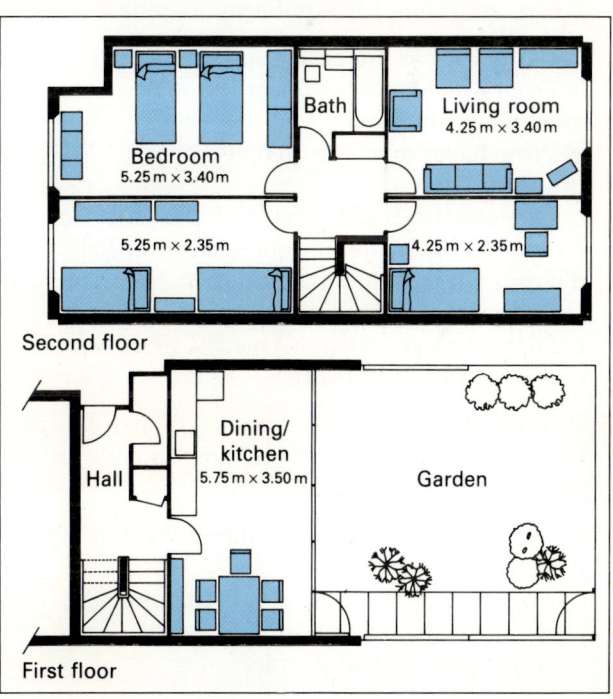

Bath

Living room
4.25 m × 3.40 m

Bedroom
5.25 m × 3.40 m

5.25 m × 2.35 m   4.25 m × 2.35 m

Second floor

Dining/
kitchen
5.75 m × 3.50 m

Hall   Garden

First floor

| | Proportion of households owning | | | | Average number of cars per 100 households |
	No car	1 car	2 cars	3 or more cars	
Runcorn	42	50	7	1	66
Cheshire	33	48	16	3	90
England and Wales	39	45	14	2	80

◀ **Figure 8.24** Table showing car ownership patterns

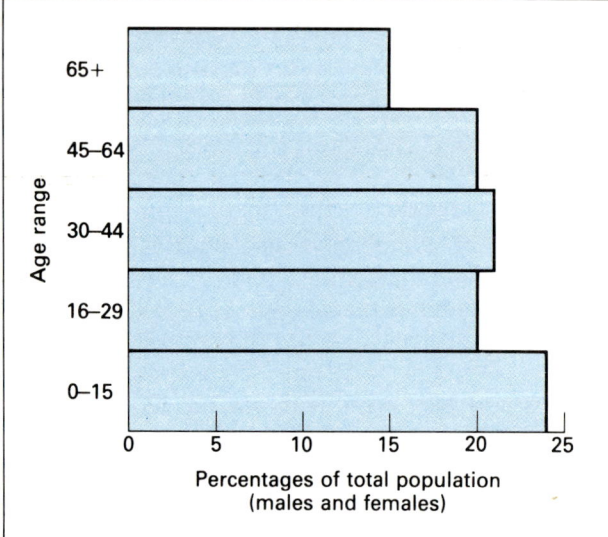

▲ **Figure 8.25** Population pyramid for the county of Cheshire

1 What are the main attractions of Runcorn New Town, according to its publicity material?

2 These questions are based on Fig. 8.18.
a From which *city* has Runcorn New Town received most migrants?
b Why have so many people come to Runcorn from this city?
c Between 1982 and 1984, did this city's contribution fall by one-tenth, one-quarter or half?
d From which *county* has Runcorn New Town received most migrants?
e Suggest a reason why so many Cheshire families choose to live in Runcorn.

3 Pair up the five listed types of owner-occupied housing in Runcorn with the most suitable 'buyer-descriptions' given below:

☐ 'growing' families, with young children
☐ highly paid executives
☐ married couples nearing retirement age
☐ newly-weds and first-time buyers
☐ single professional people.

4 **a** What is meant by the term, 'a balanced population'?
b Think of reasons why Runcorn's planners wanted their town to have a balanced population.
c How do Runcorn's home ownership/rental patterns compare with those for the whole of Cheshire (see Fig. 8.21)?
d Suggest reasons why very few of the flats in Southgate Estate have been bought by the families living in them.

5 What information in this unit proves that:
a *On average*, Cheshire people are wealthier than those throughout the country?
b Many migrants have moved to Runcorn New Town to obtain employment?

6 **a** Draw a population pyramid (similar to the one in Fig. 8.25) to show this 1981 census information for Runcorn New Town:

Age range	Percentage of population
0–15	33
16–29	26
30–44	21
45–65	12
over 65	8

b Compare the population structures shown by the two pyramids. This is best done by considering each of these age ranges separately:

☐ young people (0–19 years)
☐ young/middle-aged adults (20–60 years)
☐ older adults (aged 61 and over).

c Suggest reasons for your answers to part b of this question.

9

Perception

9.1 Introduction: the importance of perception

Perception is the way in which people think of an area. Put another way, it is the images (pictures) of it which their minds have built up. Perceptions are of course influenced by the knowledge they have acquired over the years; also by their experience of life and particularly the areas in which they are living. For example, people who have spent their entire lives in cities are likely to have perceptions which are quite different to those held by country-dwellers.

The 'media' have had a great effect on our perceptions of other, contrasting areas. Television, radio and newspapers constantly bombard us with a mass of information about the world as it is today and how it has changed over the centuries. Improved transport facilities too have played a key role in increasing our knowledge of places far and near.

Perceptions are important because they help us to make decisions. Thousands of families choose Devon and Cornwall for their holidays because they *perceive* it as being attractive – their knowledge is probably based on pictures of small fishing ports and sandy coves. Similarly, companies and even governments base their decisions on how they perceive an area or a particular problem. The questions in this unit test the accuracy of your perceptions of one urban and one rural area in Britain. The results should be quite interesting!

1 **a** Draw, *completely from memory*, a simple map of a small, well-known area in the centre of your own/a nearby town. You should name the roads on it and label any key features (e.g. public buildings, large shops, hotels).
 b Compare your own map with an 'official' map of the same area, then state any important differences between them which you have noticed.

2 Write down your own perceptions of the Lake District National Park in Cumbria. This must be done without any reference to books, maps or photographs, and may be divided up under these five headings:

Aims What are the main aims of our National Parks (i.e. reasons for having them)?

Apperance What are the chief 'physical' characteristics of the Lake District National Park (e.g. relief, lakes)?

Climate What is the Park's climate like (i.e. both temperature and rainfall)? You may use general terms such as 'warm' and 'wet' to do this.

Tourist facilities What are the Park's main facilities?

Employment How do the Park's residents earn a living (e.g. in tourism, industry)?

▲ **Figure 9.1** Our perceptions of other places aren't always as accurate as we would like them to be

9.2 Perception of environmental quality

Unit 2.7 invited you to carry out a study of the 'physical condition' of housing in your local area. Housing is of course a vital part of the urban landscape, but it is by no means the only one. This unit aims to help you to perceive and then assess the environmental quality of both urban and rural landscapes.

1 a Re-arrange the following urban environmental factors into a list showing their relative importance, starting with the most important one. There is no 'right' answer – it is a matter of personal judgement.

air pollution
derelict land
garage provision
'greenery' (e.g. trees)
indoor recreational
  facilities
industrial activity

litter and rubbish
noise
private gardens
public recreational space
  (e.g. parks)
traffic congestion.

b Give each factor in your ordered list a maximum score to show its importance; these scores will be used for the practical work in part c. It is recommended that your maximum scores should lie between 3 and 10.

c Select 2 to 5 streets taken from different parts of your local area (i.e. don't choose them all from the same residential zone!). Now carry out a survey of each street in turn, carefully recording the scores which you feel are appropriate.

d Present the information you have gathered in the form of a table, which could have a layout like the one shown below. The last column with have to be repeated for each additional street.

Urban environmental factors in rank order	Maximum score	Score awarded to (name of street)
Totals:		

▼**Figure 9.2** Drawing for use with Question 1

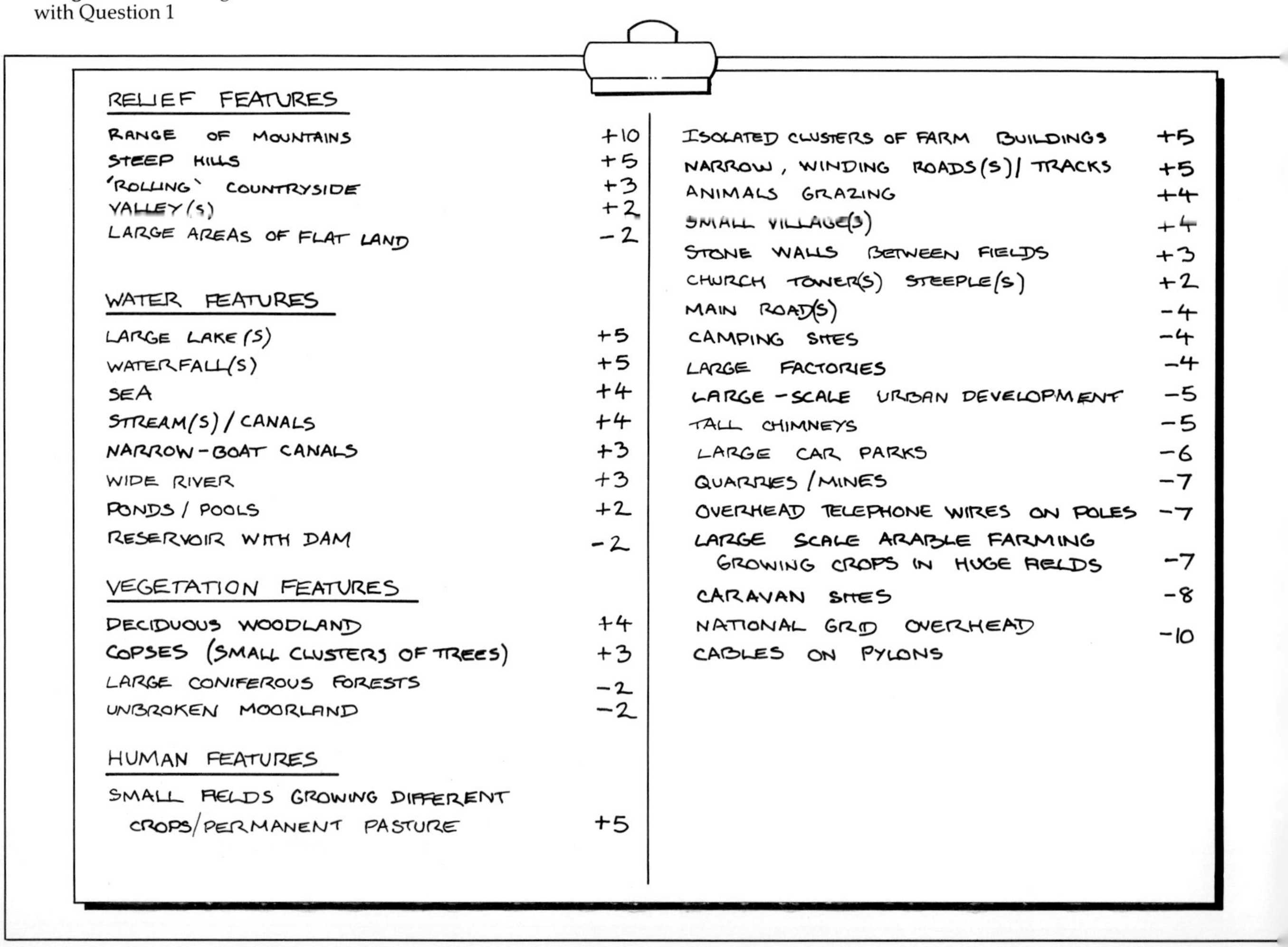

RELIEF FEATURES

RANGE OF MOUNTAINS	+10
STEEP HILLS	+5
'ROLLING' COUNTRYSIDE	+3
VALLEY(S)	+2
LARGE AREAS OF FLAT LAND	−2

WATER FEATURES

LARGE LAKE(S)	+5
WATERFALL(S)	+5
SEA	+4
STREAM(S) / CANALS	+4
NARROW-BOAT CANALS	+3
WIDE RIVER	+3
PONDS / POOLS	+2
RESERVOIR WITH DAM	−2

VEGETATION FEATURES

DECIDUOUS WOODLAND	+4
COPSES (SMALL CLUSTERS OF TREES)	+3
LARGE CONIFEROUS FORESTS	−2
UNBROKEN MOORLAND	−2

HUMAN FEATURES

SMALL FIELDS GROWING DIFFERENT CROPS/PERMANENT PASTURE	+5

ISOLATED CLUSTERS OF FARM BUILDINGS	+5
NARROW, WINDING ROADS(S)/ TRACKS	+5
ANIMALS GRAZING	+4
SMALL VILLAGE(S)	+4
STONE WALLS BETWEEN FIELDS	+3
CHURCH TOWER(S) STEEPLE(S)	+2
MAIN ROAD(S)	−4
CAMPING SITES	−4
LARGE FACTORIES	−4
LARGE-SCALE URBAN DEVELOPMENT	−5
TALL CHIMNEYS	−5
LARGE CAR PARKS	−6
QUARRIES / MINES	−7
OVERHEAD TELEPHONE WIRES ON POLES	−7
LARGE SCALE ARABLE FARMING GROWING CROPS IN HUGE FIELDS	−7
CARAVAN SITES	−8
NATIONAL GRID OVERHEAD CABLES ON PYLONS	−10

▲**Figure 9.3** Generally attractive rural scene containing some of the 'negative aspects' listed in Fig. 9.2.

e Make a detailed assessment of the quality of life in your local area, based on the contents of your table. Add comments to *explain* your findings whenever possible. After doing this question, what changes – if any – would you make:

☐ To the *rank order* in which you arranged the 11 urban environmental factors?
☐ To the *maximum scores* you gave to these factors?

Giving reasons for any amendments you feel are necessary.

You will have your own perceptions of 'beautiful' *rural* landscapes, but would probably find it quite difficult to assess their scenic value accurately. The scoring-system in Fig. 9.2 is just one way of doing this, and you might wish to invent a 'personal' system which suits you better!

2 Study the landscape in Fig. 9.3 very carefully, then use the scoring-system in Fig. 9.2 to assess its scenic value.

3 In *your* opinion, what are:
  a The two most attractive components in this landscape?
  b Its two *least* attractive aspects?

4 Make any amendments you now consider necessary to the scoring-system recommended in this unit. You can:

☐ Delete any item.
☐ Change any score.
☐ Add any extra item, in which case you must also suggest an appropriate score.

5 Make an accurate scenic assessment of a rural landscape of your own choice; it can be local, or in another area you know well. You may use: *either* the scoring-system you amended in Question 4, *or* another system which you have invented.

9.3 The Lake District National Park

This unit is divided up under the five headings used for Question 2 in Unit 9.1.

Aims

The Lake District is one of ten **National Parks** in England and Wales (see Fig. 10.9 on page 130). They were created between 1951 and 1953 and given this name, although it is quite misleading. *National* suggests that all the land in them belongs to the government; in fact, about 80% of their total area is privately owned (see Fig. 10.11 page 131). *Park* is also mislead-ing because it is more generally used for public open spaces in urban areas with beautifully kept lawns and gardens. Our National Parks are quite different. They are large enough to include farms, villages and even towns, and their landscape is much more natural.

National Parks are *not* a British invention! Many parks in Africa and North America are much older. The world-famous Yellowstone National Park in the United States was established as early as 1872. It was not until 1949 that the British parliament passed a law to create National Parks in this country. Figure 9.4 illustrates the main aims of our own parks and the kinds of activity it was hoped they would encourage.

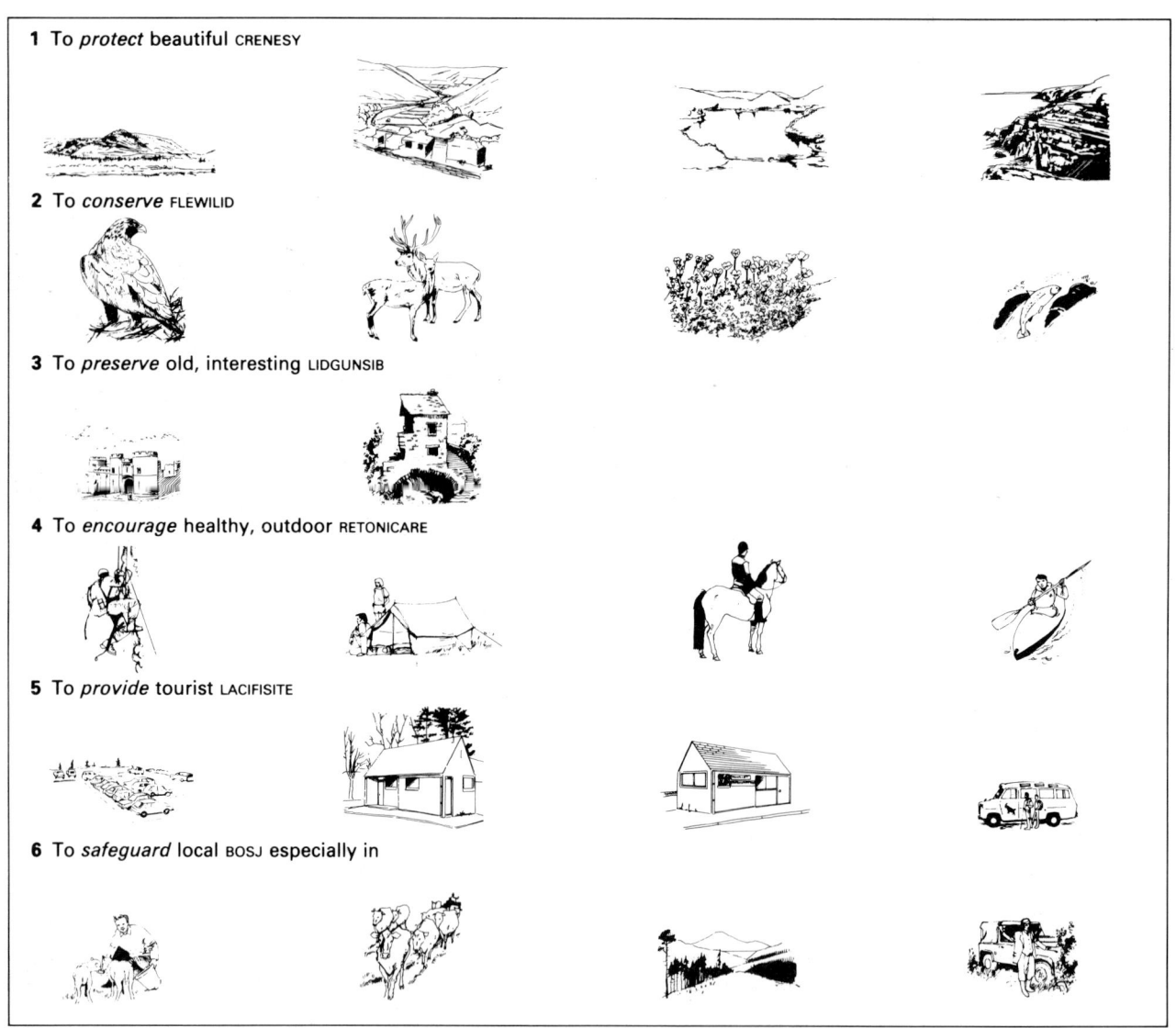

1 To *protect* beautiful CRENESY

2 To *conserve* FLEWILID

3 To *preserve* old, interesting LIDGUNSIB

4 To *encourage* healthy, outdoor RETONICARE

5 To *provide* tourist LACIFISITE

6 To *safeguard* local BOSJ especially in

▲**Figure 9.4** The main aims of the British National Parks. To find out what they are, you need to un-jumble the words in capital letters

Figure 9.5 Map showing the Lake District's three main rock areas

Key

A Slates
B Volcanic rocks
C Shale and limestone

— Rock-type boundary line
--- National Park boundary
Lake
▲ Highest point within rock-type area

Appearance

The Lake District is Britain's largest and best-known National Park. Its 2280 km² includes Windermere, our largest lake, and Scafell (978 m) – England's highest mountain. One of the main attractions of the Lake District is the great variety of scenery within an area only 65 km across at its widest point. Three types of rock cover most of the National Park area (Fig. 9.5) and they have produced the contrasting scenery shown in Figs 9.6, 9.7 and 9.8. Notice that the height of the land varies a great deal, and that the sky-line can be either jagged or smooth.

◀ Figure 9.6 Wastwater with Great Gable

▲ Figure 9.7 Skiddaw

▼ Figure 9.8 Fell Foot Country Park at the southern tip of Windermere

About 40 million years ago, all three rock areas were forced upwards and folded into a dome-shape. This formed the Cumbrian mountains in which even today the highest parts are in the central area. The surface of the dome changed dramatically during the Ice Age. which lasted one million years and ended only 12 000 years ago. The much lower temperatures allowed masses of ice to build up on the mountain tops. This ice slithered down to fill the valleys with glaciers (rivers of ice). Ice is one-third the weight of solid rock so it was able to change the shape of the valleys by eroding (wearing away) their sides and bottoms (see Figs 9.9 and 9.10).

▲ **Figure 9.9** Stock Ghyll near Ambleside – a river valley

◀ **Figure 9.10** The Langdale valley – shaped by glaciers

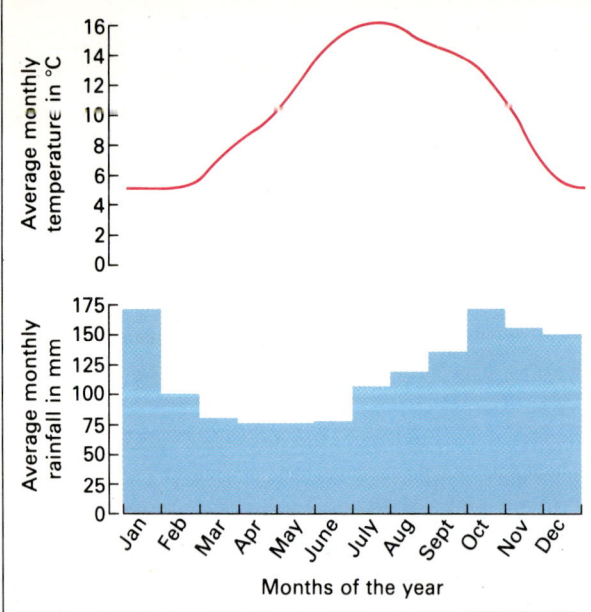

When the ice melted, it deposited (dropped) the rock and soil which it had eroded on its way to lower land. This debris would often form ridges across the valleys, which could then fill with water to produce long, narrow 'finger' lakes. Figure 9.5 shows that these lakes seem to make a 'radial' pattern around the high, central area – rather like spokes from the hub of a bicycle wheel. The appearance of this basic **glaciated** landscape has been changed by centuries of human activity (see the paragraph on employment).

Climate

The climate pattern for Keswick shown in Fig. 9.11 is fairly typical of the Lake District as a whole. There are however variations in both rainfall and temperature. The table in Fig. 9.12 indicates that the rainfall received in a typical year is linked to the height of

▲**Figure 9.11** Climate graph for Keswick, Cumbria

▶**Figure 9.12** Table to compare height and total annual rainfall from west to east across the Lake District

Distance from Irish Sea (km)	0	10	20	30	40	50	60
Height (m)	0	300	900	600	250	400	200
Total annual rainfall (mm)	625	2 000	3 375	3 200	2 175	1 500	925

◀ Figure 9.13 Scenes from the Lake District

the surrounding land *and* its distance from the west coast. Penrith, for example, is in the rain-shadow of the Cumbrian Mountains, which means that it is sheltered by them from the prevailing (almost constant) south-westerly winds. The height of the land is responsible for much of the rainfall in the Lake District. Rainfall produced in this way is called orographic (relief) rainfall. The mountains also have a cooling effect, as temperatures decrease at a rate of 1°C for every 160 metres change in height.

Tourist facilities

The Lake District was first 'discovered' by tourists in the 1700s, but it was the building of the railways in the middle of the nineteenth century which first enabled large numbers of people to visit the area. The M6 motorway has also increased the Park's accessibility.

The Lake District offers many different types of accommodation. It also has a wide range of tourist facilities. Both are illustrated in Fig. 9.13.

Employment

Tourism has created many seasonal jobs in the area. Much permanent work is also available in sheep and dairy farming, slate quarrying, craft/souvenir-making, the manufacture of pencils at Keswick, and in providing accommodation, catering services, shopping facilities and transport. Reservoir maintenance and forestry are becoming increasingly important.

1 a Produce a *detailed* account of the Lake District area under the five headings used in this unit. The following hints will help you to do this, but use your own words as far as possible.

Aims The main aims of the National Park may be listed or illustrated in simple drawings.

Appearance The following table may be used for part of you answer:

Area	Main type of rock in area	Description of rock hardness	Name of highest point	Height of highest point in m	Description of scenery
Northern					
Central					
Southern					

What are the main differences between the river valley in Fig. 9.9 and the ice-shaped valley in Fig. 9.10 (i.e. size, shape)?

Climate Describe the temperature and rainfall patterns in Keswick's climate graph (see table below). State how this pattern may change with height and distance from the coast.

Temperatures	Descriptions
Below − 10°C	very cold
− 10°C to 0°C	cold
0°C to 10°C	cool
11°C to 20°C	warm
21°C to 30°C	hot
Over 30°C	very hot

Tourist facilities List the many kinds of facility shown in Fig. 9.13.

Employment Arrange the different types of work available in three separate groups:

☐ *Primary industries* (agriculture; obtaining natural resources)
☐ *Secondary industries* (manufacturing goods)
☐ *Tertiary industries* (giving a service of some kind).

b Summarise how your perception of the Lake District National Park for Question 2 in Unit 9.1 compared with your account for a above. Include those aspects in which your perception was especially accurate or inaccurate.

2 With the help of an atlas, name the highest mountain in each of these National Parks, and add their heights in metres.

☐ Lake District National Park
☐ Peak District National Park
☐ Snowdonia National Park
☐ Yorkshire Dales National Park.

3 Copy out these statements about our National Parks, leaving out the words in italics which do not fit.
a *Few/Most* of our National Parks include stretches of coastline.
b Britain's National Parks are located in *England/and Scotland/and Wales/and Northern Ireland*.
c The parks are mainly in the *north and west/south and east* of Britain.
d They are in areas of generally *high/low* relief.
e They are in *densely/sparsely* populated areas, which means that *few/many* people live in them compared to the rest of Britain.

4 Suggest the main reasons why there are no National Parks in South-East England.

10
Conflict in Rural Areas

10.1 Introduction: conflict in recreation

Recreational areas are under increasing pressure from tourists. It is estimated that 18 million Britons are 'on the road' on a fine summer's day; more than 2 million people visited the Tower of London in 1986. The result is overcrowded roads, car parks, beaches and beauty spots. It also creates a great deal of annoyance for those who live in the worst affected areas. Residents object to the noise and pollution produced by heavy traffic; they also find it increasingly difficult and dangerous to cross roads at peak holiday times (Fig. 10.1).

Tourism is also having a serious effect on the environment. The litter left by visitors is an eyesore, but at least it can be removed. The impact of countless feet on the ground is a different matter altogether. Narrow country paths are worn down and become much wider as people stray on to their verges. Figure 10.2 shows just one of many paths in the Lake District which have suffered in this way, and much hard and expensive work is now being done to rebuild the worst affected stretches. Question 4 invites you to carry out a local study of **erosion** caused by walkers on a popular footpath.

▲ **Figure 10.2** Restoration work on a popular foot path at Tarn Hows

◀ **Figure 10.1** Trying to cross the main road in Bowness

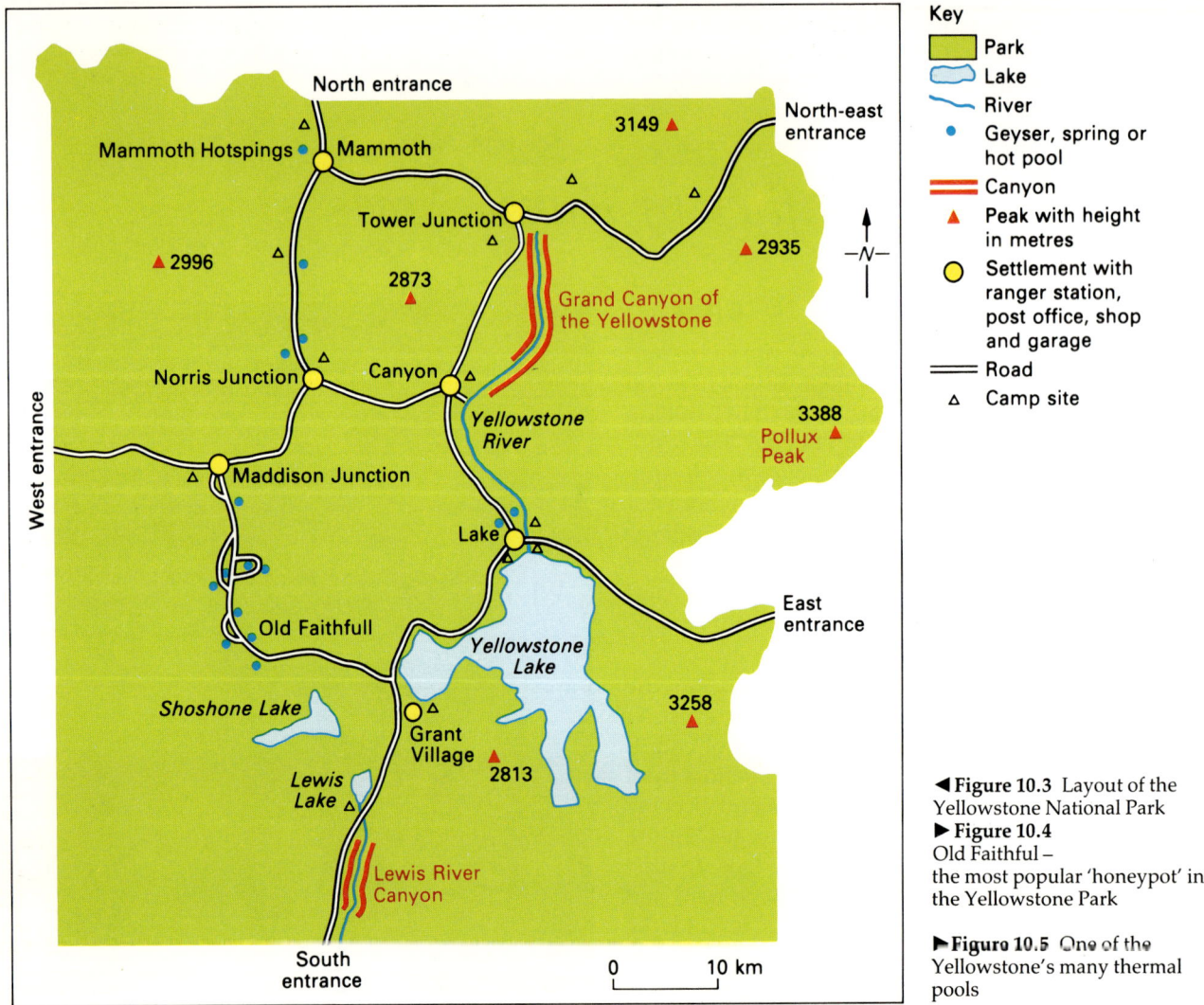

Key

- Park
- Lake
- River
- Geyser, spring or hot pool
- Canyon
- ▲ Peak with height in metres
- ○ Settlement with ranger station, post office, shop and garage
- Road
- △ Camp site

◄ **Figure 10.3** Layout of the Yellowstone National Park
▶ **Figure 10.4** Old Faithful – the most popular 'honeypot' in the Yellowstone Park

▶ **Figure 10.5** One of the Yellowstone's many thermal pools

Large-scale tourism has also produced many examples of **conflict** between different types of *recreational* activity. Fortunately, we don't all share the same leisure pursuits and this helps to spread the burden of tourism. Some people like nothing better than to visit a great cathedral or a ruined castle, while others (especially families with young children) would much rather spend their time on the beach or at a funfair. The fact that we do have so many different recreational interests means that we may not always see eye-to-eye with our fellow visitors, and the following units describe some of the conflicts which can result. For example, water-skiing is hardly compatable with swimming and nature-lovers are most unlikely to welcome the activities of the local hunt! The rest of this unit looks at recreational conflict in a national park.

The Yellowstone National Park is in the state of Wyoming, a *very* sparsely populated area in the Rocky Mountains of North America (see page 157). This area was little used by people until it was declared a National Park in 1872 – the first such park in the United States, and indeed the whole world.

The first white people to occupy the area were the 'mountain men', who made a living by trapping animals and selling their fur. They gave the Yellowstone River this name after discovering some magnificent cliffs of yellow rock on its banks. Stories of this and other wonderful natural sites led to a series of expeditions into the area. The photographs taken on one such expedition in 1871 impressed the American government so much that Congress passed a law to preserve the area for the benefit of future generations.

The park of today is a very different place to the one founded in 1872, although it is thought that only one-twentieth of its land area is heavily used by tourists. The most popular parts – the **honeypots** as they are often called – are around the best known natural features. The outstanding example is Old Faithful, a *geyser* which ejects a fountain of water high into the air regularly every 65 minutes – hence its name (Fig. 10.4). Other natural attractions include thousands of *thermal pools* – pools of hot water, or coloured mud which bubbles like thick soup about to boil over (Fig. 10.5). The geysers and pools can be very dangerous. Some children – and adults – have been killed or received first-degree burns by falling into the pools, many of which reach temperatures well over 200°C. Making the geysers and pools 'accident-proof' is a formidable task, as there are more than ten thousand of them in the Yellowstone.

The wildlife is impressive too. Vast forests of lodge-wood pine give shelter to grizzly bears, buffalo, deer, elk, coyotes and eagles. Many of the Yellowstone's larger animals such as bears and buffalo grow thin and weak during the cold winter months. Some of them die of starvation during this season of 'winter-kill'. The survivors must eat large quantities of food during the summer to make up the weight they have lost, and often visit the tourist areas in the hope of getting an easy meal (Fig. 10.6). The grizzly bears which do this are a particular worry, as their official biological name *Ursus horribilis* suggests! Usually calm tempered, the grizzly's mood can turn savage without warning and a number of children cuddling them or just offering them food have suffered dreadful injuries. The Rangers have a strict policy of returning grizzlies which become over-dependent on humans to the more remote areas of forest.

▲ **Figure 10.6** Grizzly bears often *look* friendly!

The very high numbers of vehicles now entering the Park pose a variety of problems. Traffic jams are quite common, and may reach 2 km long. These not only produce congestion on the roads but encourage unpleasant and dangerous clouds of exhaust fumes to build up on hot summer days. The Rangers have tried hard to reduce traffic flow problems and the number of motor accidents which occur each day within the Park. Speed limits are in force, even on long stretches of road where a one-way system operates. By-passes have been built alongside the most heavily used roads. Visitors are now encouraged to leave their cars near the park's five entrances and sight-see from specially designed coaches called 'tallyhos' (Fig. 10.7). Extra car parks have been built to accommodate motorists willing to camp overnight, but these are costly and tend to spoil the scenery unless they can be screened from view.

The behaviour of some of the visitors has also caused serious problems in recent years. Camping is one of the most popular tourist activities in the Yellowstone, and this has attracted large numbers of young people into the park. Gangs of youths – many on motorbikes – roam the Park causing annoyance to family groups. Thumbing for lifts is another unwelcome activity. The younger visitors increase the risk of fire through carelessness, particularly in areas well away from the main roads. The rubbish left by visitors adds up to *thousands* of tonnes each year! The Rangers need to have very special personal qualities – especially tact – to deal with all these problems. Newly appointed Rangers are given an intensive training course which includes life saving, fire fighting, traffic control, drug control and even judo and karate.

Much public money has been invested in the Yellowstone National Park in recent years to provide better roads, overnight accommodation and tourist information services. The modern visitor expects to be able to drive around the park and stay in cabins or at camp sites which have excellent toilet and other basic facilities. He/she also has a seemingly endless thirst for knowledge about the Yellowstone's physical features and wildlife. Many displays, exhibitions, lectures and guided tours are provided by the Rangers. They enjoy these creative aspects of their work but are beginning to feel the pressures they put on them. Most visitors wish to buy souvenirs as presents or momentos of their stay, and there are now plenty of shops to meet this need.

The Yellowstone National Park *has* changed a great deal in recent years, but so far these 'improvements' have been limited to quite small areas and do not seem to lessen the appeal of this priceless national asset.

1 Suggest *at least* five different ways in which tourism can cause problems for *local residents*.

2 What kinds of conflict are likely to occur between the following pairs of activities:
  a Bike-scrambling and bird-watching?
  b Fishing and cabin-cruising?
  c Camping and farming?
  d Scuba-diving and oil-refining?
  e Orienteering and grouse-shooting?

◄**Figure 10.7** A 'tallyho', a bus specially designed for taking tourists around the Yellowstone National Park

►**Figure 10.8**

3 a Obtain a copy of 'The Country Code' and write down the ten rules in it which aim to reduce conflict and damage in rural areas. You could add simple drawings to illustrate these rules.
b Are there any other rules which you feel should be *added* to the code?

4 a Carry out a survey of the effects of foot erosion on a heavily used path in your local area; 1:50 000 scale Ordnance Survey maps will show you the routes of public footpaths near to your school or home. You could work in pairs to do this, each pair taking responsibility for one section of the path. It is necessary for you to record the state of the ground at 1/3 m (33 cm) intervals *across* the path. (A tape measure or length of string knotted at these intervals will be helpful). The ground state at each interval along the transect may be assessed according to the following scale:

A Vegetation hardly disturbed; includes wild flowers as well as grasses.
B Some disturbance evident, up to 25% of the topsoil is exposed; no flowers – only grasses.
C 25–50% of the topsoil is exposed; grasses very short and compressed by walkers.
D 50–75% of the topsoil is exposed; very compact vegetation.
E Over 75% of the topsoil is exposed; very compact vegetation.

b Figure 10.8 gives one way of showing the information gathered along the transect.
c Measure the distances across the transect which are:

☐ affected by foot erosion (distance 1)
☐ most seriously affected by foot erosion (distance 2).

d This process can be repeated a number of times along your stretch of path, and the averages for each type of distance obtained.
e Record any natural or human-made factors which may have encouraged or restricted foot erosion along your stretch of path (e.g. steepness of verges, presence of fences or hedges, outcrops of rock on path or verges).
f Comment on your findings. You could mention any likely reasons for the path's heavy use (e.g. horse-riding, walking dogs, nearness to residential areas, quality of scenery). You could also suggest any ways in which these techniques might be improved.
g Your teacher may wish to collate the information obtained by the whole group and plot this on a large-scale map of the surveyed stretch of the path.

5 What do you consider to be the most likely harmful effects of large numbers of visitors on tourist 'honeypots' such as the Tower of London *and* their surrounding areas (see page 125)?

6 After class discussion, if necessary, answer the following four questions about conflict within the Yellowstone National Park. You should use the information given in this unit to support your answers.

Should visitors be allowed to come into close contact with the larger animals in the park?

Should all the geysers and hot pools in the Yellowstone be fenced off for safety reasons?

Should the number of private vehicles entering the park be strictly limited, and if so, how could this be done?

What special measures need to be taken to reduce the problems caused by young people in the Yellowstone?

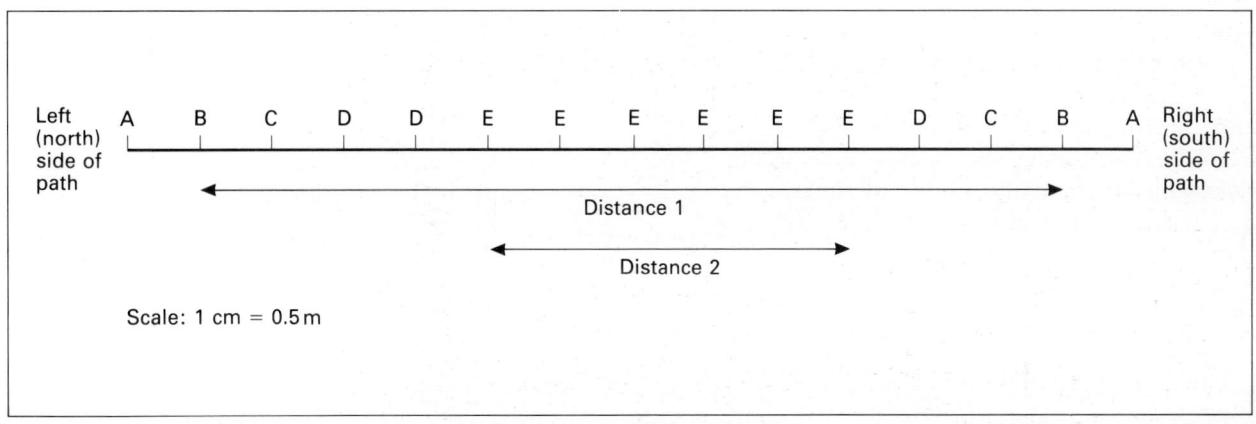

10.2 *Accessibility of recreational areas*

Accessibility describes the ease with which people can visit a place. It is concerned not only with the travelling involved in getting to a destination but also the right to visit property there which belongs to other people.

Study Fig. 10.9, which shows the layout of the motorway network in England and Wales. The main purpose of the motorways is to improve road links between our chief ports and industrial areas. They have however also proved very useful to tourists wishing to travel long distances. Only a few motorways were built especially for tourist traffic; one such motorway is the M55, which links Blackpool to the M6.

Key

— Main motorways
• Important cities
▨ National Parks. These are:
1 Northumberland
2 Lake District
3 Yorkshire Dales
4 North Yorkshire Moors
5 Snowdonia
6 Peak District
7 Pembrokeshire Coast Path
8 Brecon Beacons
9 Exmoor
10 Dartmoor

0   100 km

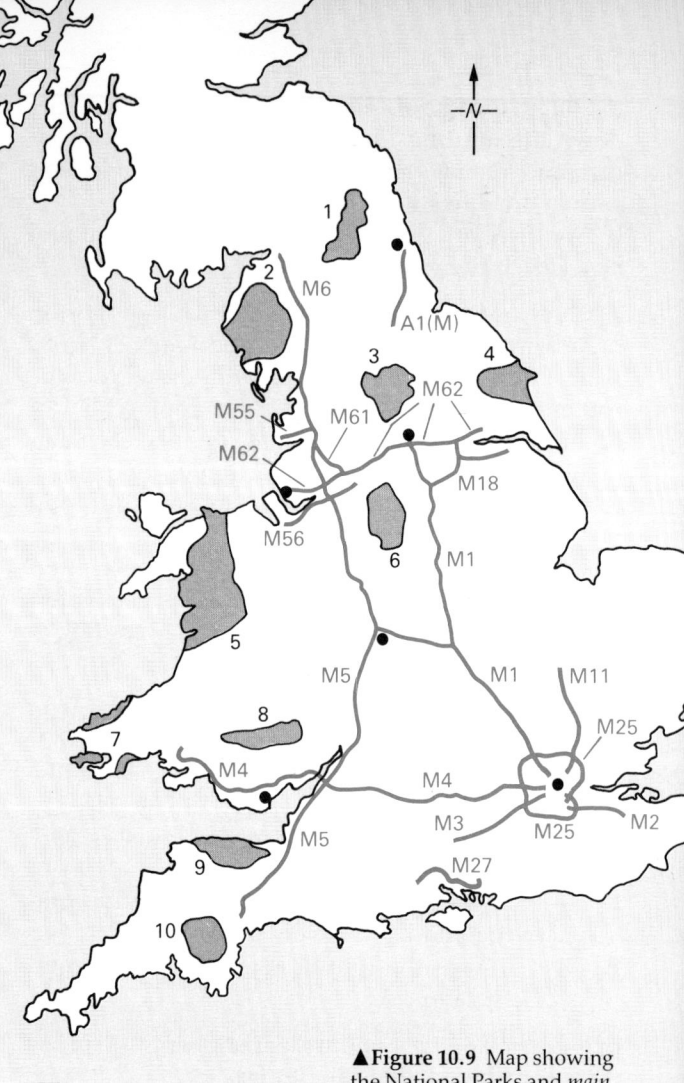

▲**Figure 10.9** Map showing the National Parks and *main* motorways in England and Wales

Many motorways pass quite close to rural recreational areas and have made them much more accessible. The more remote National Parks such as Dartmoor and the Lake District have benefited most, as millions of people in our industrial regions can now reach them much more quickly. For example, day-trippers from Birmingham can drive to either of these parks in the morning, enjoy a few hours there, and return home later in the evening. The worked example on this page shows how travelling times between Birmingham and Windermere have decreased as a result of the building of the M6 motorway.

The pre-motorway return journey

Approximately 510 km along ordinary roads, at an average speed of 60 kilometres per hour.

$$\text{Time taken} = \frac{510 \text{ (km)}}{60 \text{ (kph)}} = 8\tfrac{1}{2} \text{ hours}$$

Today's return journey

Approximately 440 km along motorways at an average speed of 110 kph *plus* 70 km along ordinary roads at an average speed of 70 kph. The average speed along the ordinary roads has been increased by 10 kph because these roads have improved considerably over the last thirty years.

$$\text{Time taken} = \frac{440}{110} \text{ plus } \frac{70}{70} = 4 \text{ plus } 1 = 5 \text{ hours}$$

The reduction in travelling time for the return journey is $8\tfrac{1}{2} - 5 = 3\tfrac{1}{2}$ *hours.* This is a considerable saving, which gives extra free time within the park itself. The reduced travelling time of 5 hours is acceptable for the occasional 'distance trip', especially as driving on motorways usually avoids traffic jams and is far less frustrating than using conventional roads.

▲**Figure 10.10** Some people don't want the general public to cross their land

▼**Figure 10.11** Divided bar graph showing land ownership in Britain's ten National Parks

Travel by public transport

Travelling times by rail have also decreased sharply over the last thirty years, but this improvement has had less effect on rural recreational areas. Rail and coach services are still used for long distance travel to resorts for the main annual holiday, but cars are the most popular form of transport for touring rural areas.

The question of accessibility to *property* is a very different matter to travelling on our public roads. Most of the land and buildings in this country are private property, and it is up to the individual owner to decide whether to allow access by the general public. Even our 'National Parks' are largely private property (Fig. 10.11).

Understandably, owners are cautious about inviting the public on to their land. They fear that the visitors will damage their property through thoughtless behaviour or by the constant impact of feet on it. There are the possibilities of fire and theft, and litter is another problem which has to be seriously considered.

Fortunately, many landowners *do* allow access, but this was not always the case. The process of opening up the countryside for recreational use has been a long and often stormy one. In 1810, the poet William

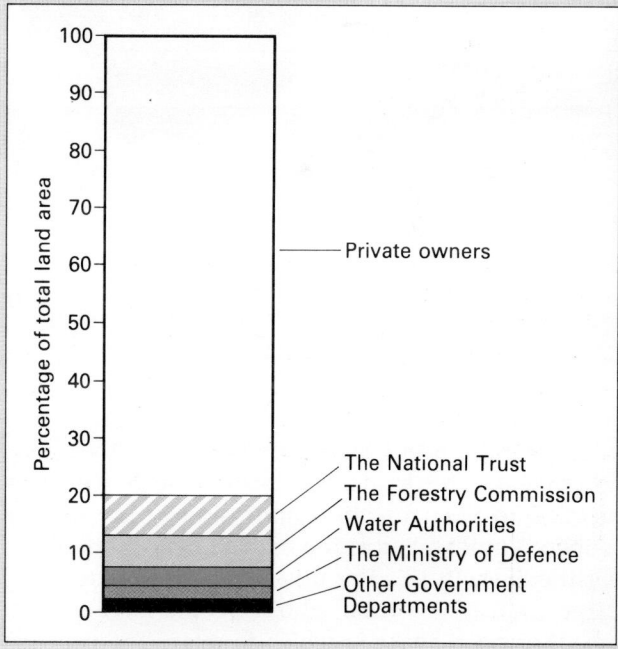

Wordsworth declared that, in his opinion, the whole of the Lake District should 'belong to the nation'. However another Lakeland poet, John Ruskin, said that he didn't wish 'ordinary working people to see Helvellyn (mountain) while they are drunk'!

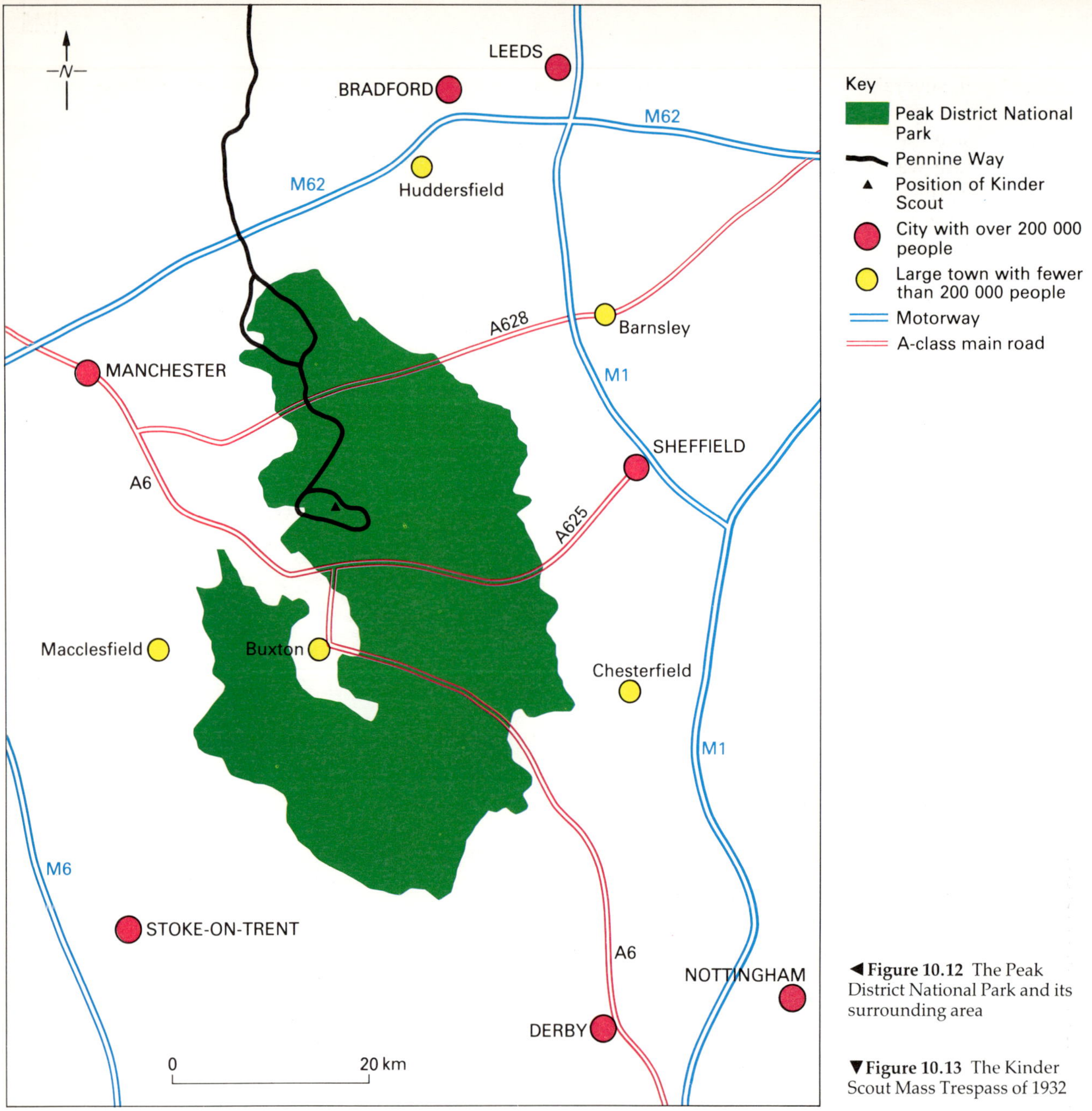

Key

Peak District National Park

Pennine Way

▲ Position of Kinder Scout

● City with over 200 000 people

● Large town with fewer than 200 000 people

Motorway

A-class main road

LEEDS

BRADFORD

M62

Huddersfield

M62

A628

Barnsley

MANCHESTER

M1

A6

SHEFFIELD

A625

Macclesfield

Buxton

Chesterfield

M1

M6

STOKE-ON-TRENT

A6

NOTTINGHAM

DERBY

0   20 km

◀ **Figure 10.12** The Peak District National Park and its surrounding area

▼ **Figure 10.13** The Kinder Scout Mass Trespass of 1932

The Kinder Scout Mass Trespass of 1932 is a good example of the kind of pressure which has been needed to open up the countryside to the general public. Kinder Scout is the highest point in the Peak District (Fig. 10.12). This upland area is close to many large industrial towns, whose people have a great need for open space in which to relax. The problem facing these people was that the area around Kinder Scout is ideal for grouse shooting. Its wealthy landowners were concerned that visitors would disturb the grouse and reduce their income from shooting rights.

On 24 April, several hundred ramblers deliberately invaded a grouse moor on Kinder Scout. They hoped this would force the landowners to give them access in future. There was a scuffle between some of the ramblers and the local game-keepers and police. Some of the ring-leaders were arrested and sent to prison for assault and riotous assembly! The Mass Trespass didn't open up the land immediately as the ramblers had hoped. But it *did* make people think much more seriously about the problem of public access to private land.

1 Copy this paragraph then fill in the blanks in it.

'Motorways have made Britain's rural recreational areas much more a _____. This means that travelling to them by road is now much easier and quicker. Most of our motorways link important _____s and _____ areas. The M __ was however built especially to take t_____ traffic between _____ and the M __. The M6 has reduced the return journey between Birmingham and Windermere to __ hours; this is a saving of __ hours.'

2 a On an outline map of England and Wales, draw the information about National Parks, motorways and large cities shown in Fig. 10.9.
b With the help of an atlas, label the six cities on the map with their names. They are:

Birmingham
Cardiff
Leeds
Liverpool
London
Newcastle-upon-Tyne

c Draw a circle around each of these six cities with a radius which stands for 150 km.
d Write down how many National Parks lie (partly and completely) within 150 km of each city.
e Name the city which is remote from all the main National Park areas.

3 Identify the motorways and National Parks shown by each of the maps in Fig. 10.14.

4 Discuss in class why cars are the most popular form of transport for visiting rural recreational areas, then write down your conclusions.

5 Use the average speeds given in the worked example to calculated the time saved in the return road journey between Liverpool and Cirencester, which is in the Cotswold Hills. The road distances to use for this exercise are:

Pre-motorway journey 200 km on ordinary roads.
Today's journey 200 km on motorways *plus* 35 km on ordinary roads.

6 Imagine you were one of the newspaper reporters present at the Kinder Scout Mass Trespass. Write a dramatic article for your paper which will help your readers to understand what happened on that day. Your article should start with a short, crisp headline.

7 a Put this set of questions to at least ten landowners in your local area who *already allow the general public to cross their land*:
1 What are the main problems with allowing the general public to have access to your land?
2 Do you place any restrictions on *where* they may walk over your land?
3 If so, what are these restrictions?
4 Would you prefer *not* to allow the general public onto your land, if this could be easily arranged?

Don't forget to thank the people you interviewed for their time and information.

b Use a variety of techniques to display the information you have received (e.g. graphs, tables and written statements).
c Summarise the patterns of information shown in your illustrations, then add written comments to assess their implications for both the land owners and the general public in your local area.

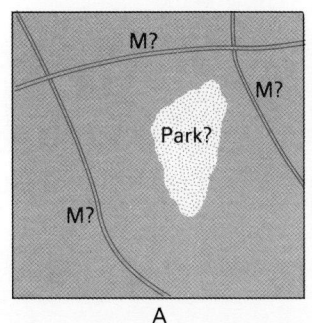
A
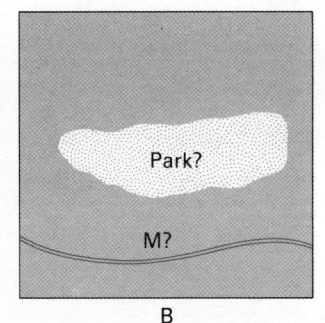
B

◄Figure 10.14

10.3 Conflict: farming

The Government has taken a much greater interest in British agriculture since the Second World War. It has approved a wide range of grants designed to make the industry more efficient and more porfitable. The EC's Common Agricultural Policy has continued and extended this level of support, and agriculture has now received more assistance than any other British industry. The aim has been to meet the cities' increasing demands for food while keeping expensive imports down to a minimum.

▼ Figure 10.15

Type of information	1938	1981
Agricultural land use (percentages)		
Arable	30	48
Pasture/rough-grazing	68	49
Other	2	3
Livestock (millions)		
Cattle	8.6	13.4
Sheep	25.0	31.4
Pigs	4.5	7.8
Poultry	76.0	135.1
Crop yields (tonnes per hectare)		
Wheat	2.2	5.8
Barley	2.0	4.3
Oats	1.9	3.9
Potatoes	17.5	28.5
Sugar beet	16.5	35.3
Workforce (thousands)	825	348
Machinery (thousands)		
Tractors	101	419
Combine harvesters	15	48
Farm size (percentage of farms)		
under 2 hectares	16	15.1
2–19 hectares	50	37.3
20–39 hectares	16	19.9
40–120 hectares	15	17.7
over 120 hectares	3	10.0

The effects of this on-going investment have been quite dramatic (Fig. 10.15); they have also included a certain amount of conflict, as new farming methods have changed the rural landscape. The much-loved 'traditional' scene has almost disappeared in certain lowland areas, especially in eastern England where the relief and the climate favour large-scale arable (crop) farming. The type of landscape which is replacing it is similar to the vast prairies of the North American plains. Although superbly efficient producers of cereal crops, it is somewhat monotonous to look at and has fewer of the habitats which conservationists know are vital to wildlife.

While many people are saddened by the loss of scenery, they are now increasingly concerned at some 'unseen' side-effects of modern farming methods. Today's farmers have about four *thousand* different chemical fertilisers and pesticides to choose from, and they now use them more heavily every year. In the past they relied on 'crop rotation' to keep the soil fertile, knowing that certain root crops would replace the goodness taken by the cereals. Pesticides certainly help the farmer to cut losses due to plant diseases, but they too can harm the environment unless used with great care. Government inspectors regularly take samples from fields and streams to check their **pollution** levels. Most of these tests reveal an 'acceptable' level of pollution. Unfortunately, chemicals which are fairly safe to use *individually* may combine to produce others which are far more toxic (dangerous). It is this 'cocktail effect' which is causing the greatest concern at the moment.

There have always been some farmers who refuse to put profit before everything else, and their number appears to be growing. One reason is that the Common Market has produced embarrassingly large surpluses of certain types of food in recent years. In 1985 alone, Britain stockpiled £1.25 billion worth of cereals, butter, beef and skimmed milk powder. This means that some European governments are now able to give extra help to those farmers who respect the environment. Another reasons is the growing strength of the 'Green Movement' and the political pressure which its conservationists can put on governments. This is now forcing them to take environmental issues much more seriously (Fig. 10.16).

1 What has been done to make European farmers more efficient producers of food?

2 Summarize the trends shown in Fig. 10.15. This is best done by writing a sentence about each of the six sets of information and quoting *some* figures to support the trends you have described.

◀**Figure 10.16** Governments now take environmental issues much more seriously

FARMING *and the* COUNTRYSIDE

COME AND SEE THEM THRIVE – *TOGETHER.*
1st to 7th June, 1986

Farmers are quietly doing a lot for the countryside. They're planting hedgerows and trees. They're creating ponds, improving woodland, and conserving moors, meadows and wetlands.

The Ministry of Agriculture, Fisheries and Food has been giving them a hand. It helps farmers with the costs of things like new hedges or footbridges and stiles. Through the Agricultural Development and Advisory Service, it gives them advice on how commercial farming can work with countryside conservation.

As well as agricultural advice, ADAS offers guidance on wildlife habitats; the effects of sprays and fertilisers; and on how things like woodlands, ponds, hedges and field borders can be looked after to benefit the countryside. It also works closely with other organisations such as the Farming and Wildlife Advisory Groups.

You can see the results of some of this work during Farming and the Countryside week, 1st to 7th June.

During the week, 33 events will be held for farmers to see conservation in practice. Many of these events will be open days on farms. The public are invited to seven of these open days. There you'll be able to see how a concern for the countryside has been made part of everyday farm management.

Ministry of Agriculture, Fisheries and Food

These seven open days will be on farms in Devon, North Yorkshire, Wiltshire, North Humberside, Gloucestershire, Warwickshire and Kent. To obtain your invitation to these public events, please contact your local office of the Ministry of Agriculture, Fisheries and Food.

FARMING AND THE COUNTRYSIDE

a week of farm events *1st to 7th* June

---

3 After class discussion, suggest ways in which each of the following are becoming increasingly worried by the high levels of subsidy (grants) which have been given to European farmers:

a Conservationists.
b The governments providing these subsidies.
c Farmers in Third World countries.

10.4 Conflict: forestry

By the end of the First World War, Britain's reserves of home-grown timber were dangerously low. In 1919, the government created the Forestry Commission and instructed it to plant fast-growing trees in 'suitable' areas. Figure 10.17 shows the location of these areas. Most of them have soil which is too thin and too acidic (sour) to grow crops. Coniferous trees such as pine and spruce were chosen, partly because they grow quickly, but also because they produce the kind of 'soft' wood which is in great demand for house construction and paper and furniture-making. Figure 10.18 shows a typical Forestry Commission development in which the trees are nearly fully grown.

Afforestation (the planting of trees on 'new' ground) has created employment in many remote rural areas. It has also led to conflict between the Commission and the general public. The earlier plantations were kept strictly out-of-bounds to visitors, because of the risk of fire. The Commission now *welcomes* people into most of its forests and provides many recreational facilities for them. Figure 10.19 shows what has been done in Grizedale Forest to make this part of the Lake District National Park more interesting and enjoyable for tourists.

▲ **Figure 10.18** Part of the Forestry Commission's coniferous plantation at Grizedale

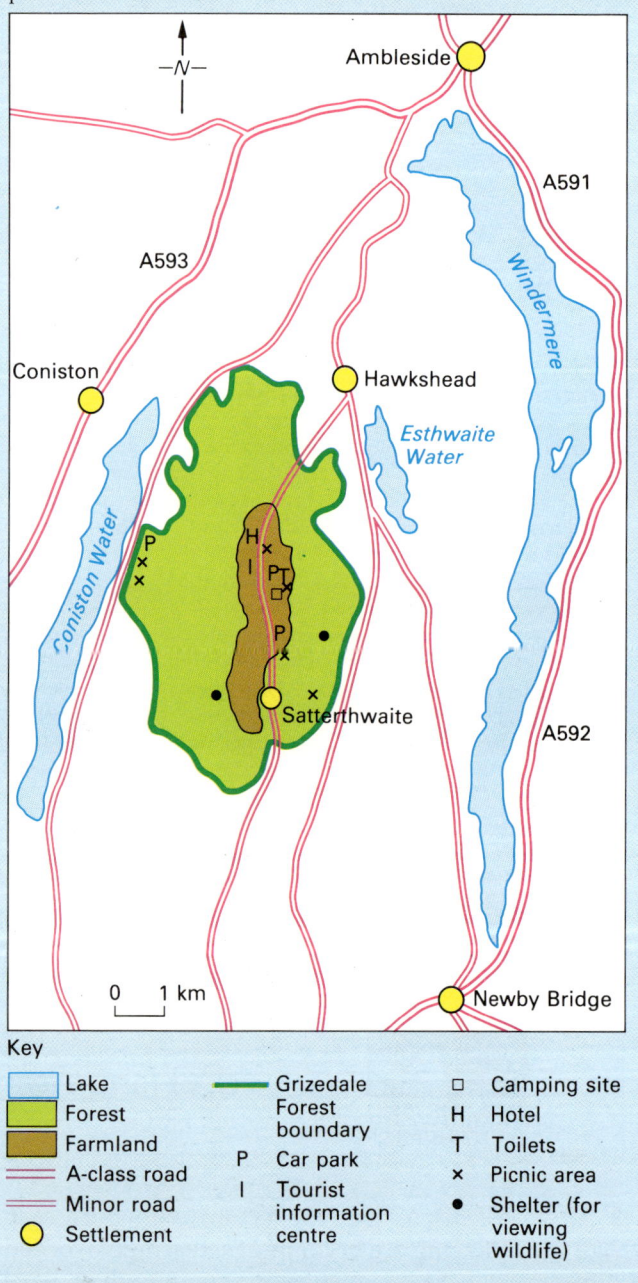

Key

Lake	Grizedale Forest boundary	□	Camping site
Forest		H	Hotel
Farmland		T	Toilets
A-class road	P Car park	×	Picnic area
Minor road	I Tourist information centre	●	Shelter (for viewing wildlife)
◯ Settlement			

▲ **Figure 10.19** Layout of Grizedale Forest in Cumbria

Key

Forest areas

Grizedale Forest

0   100 km

▲ **Figure 10.17** Britain's major forests

Conservationists have helped to bring other changes to our forests. They accept that planting fast-growing coniferous trees makes economic sense; they also know that that this type of tree quickly changes the wildlife in an area. Their permanent, dense canopy of leaves starves smaller plants of sunlight. The result is fewer species of birds and animals, and even fewer of the large creatures which prey on them.

The conservationists also resent the way in which coniferous plantations have changed the scenery; they maintain that the deciduous varieties such as oak and beech are much more attractive. The Commission has responded by adopting a more flexible policy. Many of its younger plantations are fringed with deciduous trees, giving what many people believe is a more typical 'British' type of wooded landscape (Fig. 10.20).

Conservationists are now turning their attention to *privately*-owned forests. These have been on the increase over the last 15 years, as cash grants and tax benefits have made them a very attractive investment. Today's owners include successful business-people, authors, snooker players and pop stars! Most of their new forests are of the coniferous type; they are also raising land values to such high levels that other users (e.g. sheep farmers) cannot afford to stay in business.

1 **a** Try to list at least five *traditional* uses of wood which have reduced Britain's forests over the last thousand years.
  b Now list the *present* main uses of wood, especially the soft-wood type produced by coniferous trees.

2 **a** Describe the distribution of the main forested areas shown in Fig. 10.17.
  b Suggest reasons for this distribution. Use the information provided in the text, as well as other reasons which may be obtained from an atlas (e.g. relief, temperature and rainfall).

3 Study Figs 10.18 and 10.20 very carefully, then compare the two types of scenery created by growing coniferous and deciduous trees. (Hint: consider colour, shape and variety.)

4 What recreational facilities are now available in Grizedale Forest, in the Lake District? Base your answer on Fig. 10.19.

5 Imagine that your favourite pop star (or snooker player) has decided to buy a sheep farm in the far north of Scotland and plant coniferous trees on it. Being a conservationist, you feel you must write a 'friendly but firm' letter to warn him/her of the effects of planting this type of tree on a large-scale. You should write at least 300 words.

◀ **Figure 10.20** Deciduous beach woodland in the New Forest, Hampshire

10.5 Conflict: mining and quarrying

Industry has always made great demands on rural areas which have valuable raw materials. Until quite recently, mining companies have tended to show little respect for the countryside. Raw materials were there for the taking, and the companies took a certain pride in the noise, the smoke and the turmoil which they created. The results of this attitude can be seen throughout Britain, and over 1 000 km² of our land is now classified as **derelict**. This includes abandoned buildings, worked-out quarries and tips of waste material. These are often unsightly, but can also be highly dangerous. A chilling example of this danger occurred in 1966, in the Welsh coal mining village of Aberfan. A week of heavy rain had saturated a high slag heap just behind the village, making it very unstable. On 21 August, a thick layer of the damp waste slumped (slid downwards) on to the village school, killing five teachers and 116 young children (Fig. 10.21).

Curiously, the area with the highest *proportion* of derelict land is Cornwall, a county better known for its beautiful coastal scenery and quaint fishing villages. Its dereliction has been the result of extracting two very useful raw materials: tin and china clay. Tin has been mined there for at least 2 000 years and left behind many ruined buildings (Fig. 10.22). These have mellowed with age, and some have actually become tourist attractions! The china clay workings present a very different picture (Fig. 10.23). They are

◀ **Figure 10.21** The Aberfan disaster, 21 August 1966

▼ **Figure 10.22 (left)** Ruined tin mine buildings in Cornwall

▼ **Figure 10.23 (right)** China clay workings at St Austell, Cornwall

▲ **Figure 10.24** Quarry near to Ingleborough Hill

still active and the landscape being created there is very 'raw'.

Many mineral workings are sited within our National Parks. Although these areas of exceptional natural beauty are *supposed* to be protected against unsightly developments of any kind, it is sometimes very difficult to refuse them planning permission. Raw materials are both heavy and bulky, and so it makes economic sense to extract them close to where they are needed. Sometimes the minerals are of a very high quality and the nearest alternative source may be a considerable distance away. There is also the need to create employment in any area where jobs are few and far between. The quarry shown in Fig. 10.24 is close to Ingleborough Hill – one of the outstanding natural attractions of the Yorkshire Dales National Park. This quarry continues to expand because it can produce hard chippings which are ideal for road surfacing.

Greater efforts are now being made to safeguard the environment from the worst effects of mining and quarrying. This daunting problem is being tackled in two different ways: improving the appearance of old workings with the help of subsidies from public funds, and insisting that companies make good any damage they cause in the future. Some of the waste material is being removed and used in new road-works, but this accounts for only 0.1% of the esti-

mated total of 3 000 000 000 tonnes! Much of what remains is being levelled off and then landscaped. This not only makes the tips safer and less obvious, but provides an opportunity to create additional recreational facilities.

1 What problems have resulted from mining and quarrying in the past?

2 **a** Discover for yourself what goods are manufactured from tin and china clay.
   b Write down your *own* impressions of the industrial landscapes in Figs 10.22 and 10.23.

3 **a** Explain why quarrying is sometimes allowed to take place in our National Parks.
   b According to Fig. 10.24 what harmful effects is this quarry having on the area around it?

4 In what two ways is the environment now being protected against the effects of mining and quarrying?

5 Imagine that grid reference positions 653563, 696604 and 711574 on the Ordnance Survey map on page 148 are being considered for the quarrying of building materials. With the help of class discussion, if necessary:
   a List the most likely advantages and disadvantages of quarrying at each location.
   b State which location you consider to be most suitable for such a development.
   c Give the main reasons for your choice in b.

10.6 Conflict: water resources

Water is vital to life, for no-one can survive more than a few days without it. Some countries cannot guarantee that all their people have a constant supply; this is because their climates include periods of drought *and* because they may not be able to afford to build reservoirs to store water and move it to where it is needed most. Although Britain has a less extreme climate, there have been some lengthy spells of dry weather (e.g. the summer of 1976) when we experienced water shortages.

Figure 10.25 shows that certain areas in Britain *usually* receive much less rainfall than others. These drier areas are said to be in a *rain shadow*. This simply means that the winds blowing over them for most of the year – the *prevailing winds* – have already shed most of their dampness as rain over the mountains. Figure 10.26 will help you to understand *why* there is such a close link between rainfall and relief (the height of the land). Now look back to Fig. 10.25, and see if you can spot another link – that between relief and population density. Ask your teacher for help if you find this difficult to do.

Clearly, the cities have a problem. Some of them (e.g. London) can use rainwater which has been stored naturally in layers of 'porous' rock deep below the surface. This water is very pure, but the cities have to careful not to over-use this source of supply; doing so would remove the water more quickly than the rain can replace it. Some cities (also London) can take water from nearby rivers. This water has to be purified (cleaned) before it is safe to drink. Unfortunately, these two local sources rarely match the cities' ever-increasing demand for water. The shortfall has to be met by obtaining water from areas which have a surplus. Inevitably, these are the wet, mountainous areas where fewer people live.

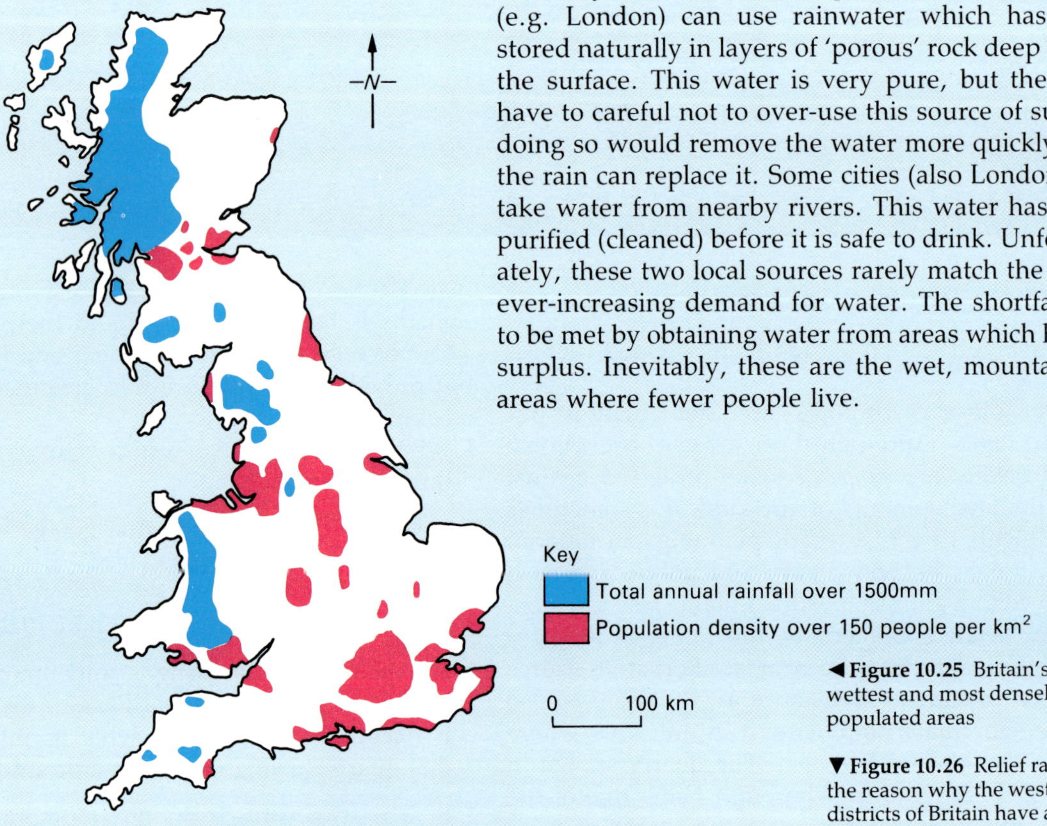

Key

■ Total annual rainfall over 1500mm

■ Population density over 150 people per km^2

0   100 km

◀ **Figure 10.25** Britain's wettest and most densely populated areas

▼ **Figure 10.26** Relief rainfall is the reason why the western districts of Britain have a particularly damp climate

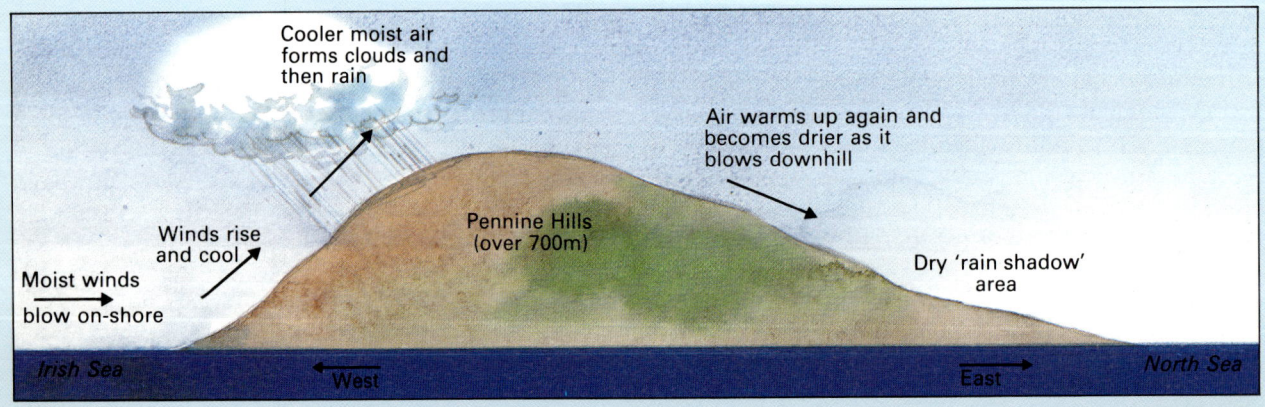

Cooler moist air forms clouds and then rain

Air warms up again and becomes drier as it blows downhill

Winds rise and cool

Pennine Hills (over 700m)

Moist winds blow on-shore

Dry 'rain shadow' area

Irish Sea   ← West   East →   *North Sea*

The rate at which water is now being taken from the upland areas is a matter of great concern to them – even though they do receive large sums of money for providing this service. Their main worry is that the water is gathered in reservoirs, which flood the valley bottoms. Lowland is scarce and precious in any mountainous region, for it provides the best sites for villages as well as the natural routeways which link them. The valleys also contain the most valuable agricultural land; their bottoms are sufficiently fertile and sheltered for grazing cattle, growing their winter feed, and protecting sheep when the fells are covered by snow. The loss of this land can make hill farming totally uneconomic. It can also force more people to leave areas which are already sparsely populated.

One alternative to flooding river valleys is to build **barrages** (long dams) across wide tidal estuaries. The barrages trap fresh water brought down by rivers, and this gradually replaces the salty sea water behind them. Possible locations for this type of development include Morecambe Bay, the Dee and Severn estuaries, Solway Firth and the Wash. All five locations have already been surveyed and could be used at a later date when inland resources can no longer meet our demand for water.

1 Pair up these key words with their correct meanings:

barrage
porous
prevailing
rain shadow
relief

– describes a layer of rock which can hold water.
– describes an area which is much drier than the mountainous areas near to it.
– describes a wind which blows from one direction for most of the year.
– describes the height of the land.
– is the name for a dam across an estuary.

2 Complete each of these statements by choosing the most appropriate word(s) in brackets.
a The wettest parts of Britain are the (highland or lowland) areas.
b The driest areas are in the (north and west or south and east).
c Most British cities are in the (highland or lowland) areas.
d (Few or most) British cities are sited by a river. (Hint: use your atlas.)
e Areas with a surplus of water have (less or more) of it than they need.

3 List separately the many arguments for and against storing water in river valleys.

4 a Construct a line graph to show this information.

Population of England and Wales (in millions)	Water used (in millions of cubic metres)
40.0	3.2
43.7	7.0
46.1	9.5
48.6	14.0

b Describe the trend in water useage suggested by your graph.
c Suggest reasons for this trend.

5 a On an outline map of Britain, locate and name the five most suitable sites for coastal barrage schemes. These are named in the text.
b After class discussion, suggest the most likely advantages and disadvantages of the barrage method of storing water.

11

Improving the Quality of Life

11.1 Introduction: the work of the National Trust

This book has described the most important changes which have taken place in rurban areas at home and abroad. The last two units study ways in which both governments *and groups of individuals* are able to influence change within these areas.

Many local and national organisations exist to protect our rural environment. For example the Friends of the Lake District aim to preserve that beautiful mountainous area for the future. The Council for the Protection of Rural England (CPRE) tries to do the same for the whole country.

The National Trust is the most effective of all these organisations because it is also the nation's largest private landowner. This means that it can influence the use and development of large tracts of our countryside and coastline. The following figures give some idea of the scale of the National Trust's operations in England and Wales (there is a separate National Trust for Scotland).

- ☐ It owns 2 400 km² of land. This is the size of a small English county.

- ☐ It *protects* a further 360 km² of land.
- ☐ It owns over 700 km of our most attractive coastline. This is much greater than the whole of the south coast of England.
- ☐ It owns all or most of the houses in 44 different villages.
- ☐ It owns more than 200 great country houses and the art treasures which they contain.
- ☐ It rents more than 1 100 farms out to its tenants.

These figures are very impressive, and are increasing every year. The work of the Trust can never be finished and the nation owes it a debt of gratitude for what it has already achieved.

The National Trust was formed in 1895 by three conservationists. They realised that some of our most attractive countryside was in increasing danger of being spoiled – by industrial growth, the lack of planning regulations to control the use of land, and the rapid increase of our population. The Government at that time recognised the value of what the Trust hoped to do and passed the National Trust Act of 1907. This important Act gave the Trust 'charity status', which means it is a non-profit making organisation and can claim some very useful tax concessions.

▲ **Figure 11.1** The National Trust emblem

▶ **Figure 11.2** Robin Hood's Bay, Yorkshire – one of the many stretches of coastline bought with funds raised by 'Operation Neptune'

▲**Figure 11.3** A caravan site on National Trust land. Notice that it is well screened by trees

▲**Figure 11.4** Sizergh Castle, north of Lancaster; although now owned by the National Trust it is still lived in by the family who have occupied it for hundreds of years

In 1937, another Act was passed which allowed the Trust to buy country houses and the treasures inside them. Many were in great danger of being sold to property developers as their owners had been crippled by heavy taxation.

The Trust raises much of the money it needs from membership subscriptions (it now has over 1 100 000 paid-up members). Extra money comes from donations, and legacies bequeathed in wills.

From time to time, the Trust launches special appeals for projects which it feels are particularly important. One such appeal is 'Enterprise Neptune', first launched in 1965. It aims to extend the Trust's ownership – and hence its protection – of stretches of unspoilt coastline. The result: a further 400 km secured for our enjoyment in the future (Fig. 11.2)

Luckily, the National Trust does not aim to turn its properties into 'fossils'. Instead, it encourages developments which are unlikely to disfigure the landscape. It knows that the scenery we enjoy today is the result of farming in the past and so it welcomes the use of caring farming methods. It also accepts that visiting tourists have certain needs and rights, and tries to meet these whenever possible. The general public are usually allowed access to Trust properties, subject to the needs of farming, forestry and the protection of wildlife.

The Trust's approach to landscape conservation is very well illustrated by its policies for wooded areas. It recognises that the natural vegetation for most of lowland Britain is deciduous woodland of oak and beech trees. These provide vital shelter for some of our most interesting birds and wild animals. They also create varied and attractive landscapes which are especially popular with tourists.

Tenant farmers on National Trust land have to obtain the Trust's approval before carrying out any 'improvements' which are likely to change the outward appearance of their property. Even lines of trees planted around their fields to act as wind-breaks need the Trust's approval! New farm buildings usually have to be built of traditional local materials. The Trust is very strict about not allowing farmers to site caravans and tents on the land they rent. The Trust's Council has now decided that new caravan sites must be located as near as possible to the boundaries of our National Parks; only small sites will be allowed on Trust land in their central areas, and only where there are enough trees to screen them from view (Fig. 11.3).

The Trust is equally sensitive about its responsibilities towards the many historic buildings which it owns. Its great country houses are often lived in by the families which used to own them. This means that they are maintained with great care and keep their warmth and atmosphere.

1 Design a poster to advertise the work of the National Trust.

2 Discover all you can about National Trust properties in your local area. Draw a map to show their location, then write a summary of any information you have collected about them.

11.2 Planning controls

Until quite recently, it was usual to build houses and factories without first obtaining planning permission. A builder simply bought some land, designed the buildings he thought people wanted to buy, then began building them. There was little control over where they were sited, or the designs used.

In 1909, the government passed the first of many laws which limited the freedom of builders. At the present time, *most* new buildings cannot be erected until the local authority has agreed that they are 'suitable'. Even *extensions* to buildings now require their approval. The main exception is farm buildings, but it seems likely that they too will be subject to planning controls in the future. *Major* developments such as motorways, airports and new towns must be approved by the government; permission is normally granted only after the general public has been given an opportunity to raise objections. This process can take a number of years, and Fig. 11.5 lists all the stages which must be followed before a new town can be built in Britain.

```
┌──────────────────────────────────────────────────────────────┐
│ Locate suitable sites                                          │
└──────────────────────────────────────────────────────────────┘
                              ↓
┌──────────────────────────────────────────────────────────────┐
│ Select the most suitable site; submit its name to the         │
│ Secretary of State for Government approval                     │
└──────────────────────────────────────────────────────────────┘
                              ↓
┌──────────────────────────────────────────────────────────────┐
│ Hold first public enquiry                                      │
└──────────────────────────────────────────────────────────────┘
                              ↓
┌──────────────────────────────────────────────────────────────┐
│ Appoint key personnel (e.g. architects) who will prepare an    │
│ outline plan for the new town                                  │
└──────────────────────────────────────────────────────────────┘
                              ↓
┌──────────────────────────────────────────────────────────────┐
│ Forward outline plan to Secretary of State for Government      │
│ approval. On receipt of approval key personnel prepare         │
│ detailed plans for the new town                                │
└──────────────────────────────────────────────────────────────┘
                              ↓
┌──────────────────────────────────────────────────────────────┐
│ Buy necessary land                                             │
└──────────────────────────────────────────────────────────────┘
                              ↓
┌──────────────────────────────────────────────────────────────┐
│ Start building the new town                                    │
└──────────────────────────────────────────────────────────────┘
```

▲ **Figure 11.5** Stages in the planning of a British New Town. County councils are resonsible for carrying them out

▲ **Figure 11.6** Demonstrations are often held to influence planning decisions. This one was against Torness Power station at Scotland

The general public are not slow to use their influence in planning matters, and many voluntary organisations have been founded to help them to do this. The National Trust, whose work was described in the last unit, is just one example. The Council for the Protection of Rural England (CPRE) is another. It has already done much to preserve some of our most attractive countryside for the benefit of future generations.

The public's involvement is not always peaceful! People are increasingly concerned about the state of their environment, and sometimes express their views very forcibly. Figure 11.6 shows a typical public demonstration; its placards leave the planners in no doubt about what they are being 'asked' to do!

1 For how many years have British builders had to obtain planning permission?

2 Who can grant planning permission for:
   a small projects such as a house extension?
   b major developments such as a motorway?

Attempt the next two questions after class discussion.

3 Suggest reasons why local residents might demonstrate against proposals to:
   a Create a dumping ground for nuclear waste?
   b Build a prison.
   c Widen a major road.
   d Extend an airport to take larger aircraft
   e Build a new reservoir.

4 Suggest any changes which you think are needed to improve the planning procedure for a new town as shown in Fig. 11.5. (Hints: Is the process too lengthy? Is there too much or too little opportunity for the public to become involved in it?)

12

Role-Play Exercises

12.1 *Urban planning enquiry*

This enquiry is based on the Ordnance Survey map extract in Fig. 12.1. Its aim is to prepare urban renewal plans for the blocks of nineteenth-century terraced housing around the primary school on Hammond Street lying within the area bounded by Ripon Street (to the north), Aqueduct Street (to the south), Green Bank Street (to the west and

Plungington Road (to the east). The area to the south of Aqueduct Street, which used to have the much older back-to-back type of dwelling, has already been re-developed. The success of this operation has prompted Preston Town Council's Planning Depart-

▼ **Figure 12.1** Enlarged section of the 1956 1:10 560 Ordnance Survey map of Preston. Crown copyright reserved

ment to look ahead to the time when the by-law housing in the enquiry area also needs to be renewed. The whole area could be *comprehensively redeveloped* or *renovated* (re-read Unit 2.5 if you have forgotten what these key terms mean). The enquiry consists of three phases.

Phase one – obtaining evidence

Flushed with their earlier success, the Council has decided to invite comments and suggestions from members of the local community and selected experts before drawing up the final plans for this area. In the first phase of the enquiry, each member of the group will be allocated one of the following roles. He/she will be asked to prepare a written report which can then be displayed on the walls of the Planning Department (the classroom). These reports need not be very long, and they may be illustrated with labelled maps and sketches. Each report should reflect the personal or 'official' views of its writer. For example, 'a mother with young children of pre-school age' is likely to insist that some open space should be provided for recreational use.

Members of the local community

1 A male teenager.
2 A female teenager.
3 A mother with young children of pre-school age.
4 A parent with older children of primary school age.
5 A retired person.
6 The owner of the post office on Plungington Road.
7 The licensee of a public house on Aqueduct Street.
8 The vicar of the church on Havelock Street.
9 The local councillor who lives in the area and represents the ward in which it is located.
10 The owner of the mill on Green Bank Street.

Experts

1 A historian who has made a special study of British housing trends during the nineteenth and early twentieth centuries.
2 A planner, who has a detailed knowledge of the British housing trends since 1920.
3 A planner, who is used to dealing with traffic and parking problems.
4 Preston's advisor for recreational facilities.
5 A specialist in current shopping trends.
6 A psychologist, who is well aware of the stresses of living in inner city areas.

7 Liverpool's Chief Planning Officer, who has been closely involved in urban renewal in Toxteth.
8 A Police Inspector, with many years' experience of policing inner city areas.
9 The Chief Education Officer for Lancashire.
10 A specialist in immigrant affairs.
11 Preston's Borough Treasurer, who is concerned that the urban renewal programme should not prove too costly for local ratepayers.
12 A representative of Age Concern – the organisation dedicated to protecting the interests of old people.

Phase two – preparation of urban renewal plans

Class members abandon their phase one roles and form discussion groups of up to four individuals. The groups will be given time to study the phase one reports on display. Each person will then be allocated one of the following duties.

1 **Chairperson** Co-ordinates the discussion of his/her own group and makes sure it is conducted in an orderly way.

2 **Housing co-ordinator** Suggests which type(s) of housing should be included in the group's plan. He/she will play a key role in deciding whether comprehensive redevelopment or renovation should be used.

3 **Facilities co-ordinator** Suggests which kinds of *amenity* should be included in the group's plan. He/she may wish to keep some of the amenities which are already in the area and/or suggest new ones for it.

4 **Road/transport co-ordinator** Suggests how the street layout in the study area should be adapted – not forgetting the need for car parking.

On completion of their discussion, each group will be invited to prepare an urban renewal plan for the area. It should consist of a carefully drawn map (complete with key) as well as a written summary of its main features and advantages.

Phase three – plan selection

In this final phase, the class should be given sufficient time to study the plans recommended by all the groups. A vote can then be taken to decide which plan most deserves to be 'submitted' to the Town Council. The teacher may wish to conclude the role-play exercise by discussing the relative merits of the groups' proposals.

12.2 Green belt planning enquiry

The following article appeared in the *Daily Telegraph* on 28 October 1985.

Green belt plan for new village

By John Grigsby, Local Government Correspondent

Proposals for a 'high-tech' new village in the metropolitan green belt at Leybourne, near Maidstone, Mid-Kent, have been put forward by Ideal Homes, the house-building arm of Trafalgar House.

The proposals, now being discussed by local councils, include 2 000 homes, a 'hightech' industrial park, a large district shopping centre, and sporting and recreational facilities, including an ice-rink, on 450 acres of land. (= 182 hectares).

The development would clearly benefit from a Channel Tunnel road link, for which Trafalgar House is among the competing consortia.

The Kent proposal is the latest of a series for large housing developments of 2 000 homes or upwards, which are described as 'new villages' because their builders propose to incorporate facilities such as open spaces, schools and community halls, which are normally provided by local councils.

At least six and possible eight are in the pipeline. The fate of all will almost certainly have to be decided by Mr Baker, the Environment Secretary, and they face him with a potentially explosive political judgment.

Farm land

The land in Mid-Kent lies just about a mile within the outer edge of the Metropolitan green belt on what local people say is good farm land.

The fear that Kent County Council, which is currently reviewing its structure plan, which sets out the development strategy for the area, may be tempted to move the belt the order to allow the development to go ahead.

Part of the site covers a psychiatric hospital, closed down under the Government's 'care in the community' proposals, and earmarked for development.

Mr Roy Barnes, vice-chairman of the planning committee on Conservative-controlled Tonbridge and Malling-

borough council, said yesterday: 'We are absolutely against the proposals. We believe that the area has already taken quite enough development recently, and that these proposals would swamp the character of the area. We shall be fighting all the way.'

Details refused

Trafalgar House refused to give details of the proposals, but it denied that the whole site lay within the green belt. A spokesman said: 'We have been having exploratory talks with the local authorities to find out what is acceptable to them.'

Mr Tim Thompson, acting planning director of Tonbridge and Malling, said: 'This is the latest in a series of challenges to the integrity of the green belt.'' Not all the other large settlements proposed in the Home Counties are on green belt land; but all are likely to provoke fierce political controversy, and will depend on the structural plan reviews.

This newspaper article shows that there is increasing pressure to build houses within 'green belt' areas. The most sought-after locations are nearest to existing rail and motorway links with the city centres. The idea of building new villages in rural areas seems very sensible; new jobs are created and the local councils' incomes from the rates are greatly increased. In the case of Leybourne, the proposed 'high-tech' village would replace many of the jobs lost when its large psychiatric hospital closed down recently.

The main objection to such schemes is that green belt areas are supposed to be protected from large-scale developments! The Leybourne site is only just inside London's green belt, and this is likely to produce a fierce debate at the planning enquiry stage. The article suggests that the council may be tempted to move the green belt boundary far enough to put the site under normal planning controls. Environmentalists will regard Leybourne as a test case. They are certain to argue that green belts are essential to the quality of life of our city-dwellers, and losing this

particular 'battle' would endanger *all* of Britain's green belts.

Leybourne is 40 km south-east of the heart of London, in an area of fertile rolling countryside. Ten years ago it was described as 'little more than a castle, a church, a large hospital, a couple of farms and some modern bungalows along London Road' (the A20). Since then at least 1 000 new houses have been built on a large estate to the south of the M20/A228 road junction. This includes a school and a modern shopping parade. The large and now vacant hospital site has attracted a good deal of interest. Trafalgar House are likely to offer about £30 million for the site – *if* they get planning permission to build their high-tech village. The Home Office believes this land could easily be adapted for use as an 'open prison', for which they would pay only £3 million. A similar proposal to use part of West Malling Air Station as a borstal failed in the early 1970s due to the strength of local opposition.

The role-play exercise is based on the information on page 147, and the maps and photographs provided in the rest of this unit. The exercise has two separate phases.

Phase one – preparation of displays

Each member of the class is asked to prepare a piece of information for display in the Planning Enquiry Room (the classroom). These displays will take the form of maps, illustrated written reports, and letters giving the local community's opinions of the proposed high-tech village development.

The maps should be drawn to the same scale, and be based on the Ordnance Survey 1:50 000 extract in Fig. 12.2. A piece of writing should accompany each map to summarise the information it is intended to

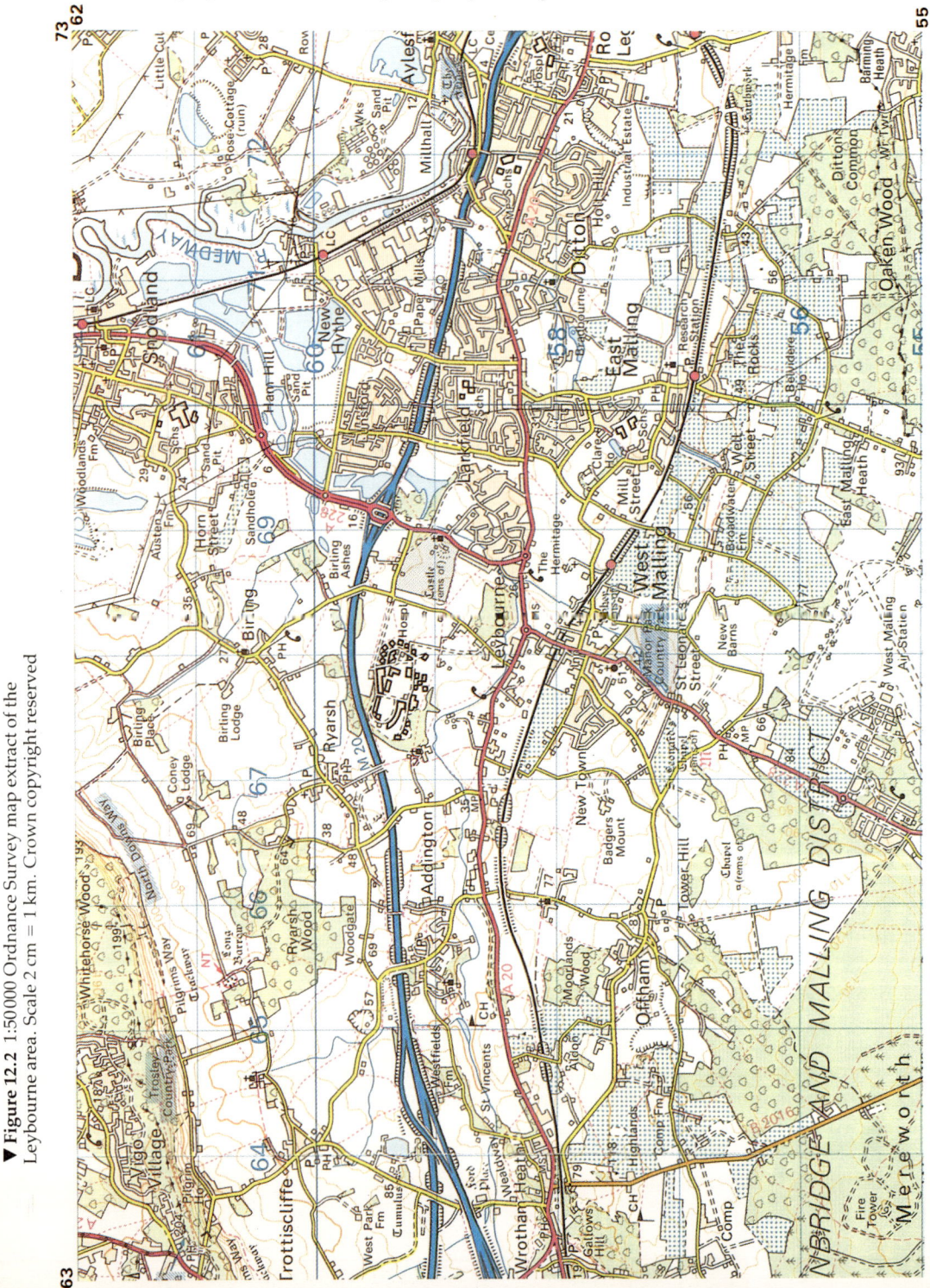

▼ **Figure 12.2** 1:50 000 Ordnance Survey map extract of the Leybourne area. Scale 2 cm = 1 km. Crown copyright reserved

show. The various maps could concentrate on:

1 Relief.
2 Distribution of main built-up areas.
3 Existing road network.
4 Existing railway network.
5 Main woods, lakes and streams.
6 Other interesting features (e.g. old buildings and recreational facilities).
7 The proposed site of the new village.

The written reports could be based on the following topics. Diagrams and labelled sketches may be added to make them more useful. Some of these topics require additional research.

1 Main types of farming around Leybourne.
2 Scenery in the North Downs and the Weald.
3 Types of work available in the Leybourne area.
4 The present layout and appearance of Leybourne village (including types of houses).

▲ **Figure 12.3** In which pink shows the area of proposed development, brown shows older housing and yellow, recent housing

► **Figure 12.4** Leybourne Castle, Kent. The castle, which was re-built by Sir Roger de Laybourne in 1190, is now part ruin and part house

► **Figure 12.5** New housing in Leybourne, Kent

► **Figure 12.6** A typical rural scene near Leybourne, Kent

The letters should be addressed to the Editor of the local newspaper, the *Kent Messenger*, begin 'Dear Sir or Madam', and end with 'Yours Faithfully'. They should be personal letters, intended to express opinions, suggestions and matters of concern to the individual writers. They could include:

1 An unemployed person who used to work at the psychiatric hospital.
2 A high-tech worker who lives in Leybourne but is travelling into London where his employers' factory is based. The journey is expensive and time-consuming.
3 A bored teenager living in Leybourne.
4 A Leybourne housewife, with children at school.
5 A local builder.
6 A local old age pensioner.
7 A local shopkeeper.
8 A local environmentalist who feels that old Leybourne and the green belt area must be protected from large-scale development.
9 A local Councillor who is in favour of the proposal to build the high-tech village.
10 A local Councillor who is very much against the proposal.

Phase two – the debate

Your teacher will allow you enough time to make a careful study of all the completed material on display. You will also need to use this time to prepare your own thoughts for the debate which follows.

The purpose of the debate is to give the 'local people' who wrote the letters an opportunity to put questions to those who prepared the map displays and written reports. They should choose questions which help to clarify the main planning issues involved – both for and against the proposal to build a high-tech village at Leybourne. After answering the questions put to them, the researchers may wish to add their own questions or state opinions which they believe to be important.

The Chairperson (the teacher) will probably insist that all questions are directed through him/her. This can be done quite simply by saying 'Mr/Madame Chairman, I would like to ask Mr/Mrs ... why/ whether ...'. This procedure is often used to make sure that lively discussion (which is to be encouraged) does not become too bitter or personal. The Chairperson may also re-arrange the furniture so that everyone has a clear view of the proceedings, and issue each pupil with a card showing his/her role-play name and character.

When all the main planning issues have been debated thoroughly, the Chairperson will summarise what has been said. He/she will then invite the letter-writers to vote on the proposal that 'Ideal Homes should be allowed to build the proposed high-tech village on the site of the old hospital'.

1 Draw a map of the area covered by Fig. 12.3, to include the different types of information which *you* consider most important in this planning issue.

2 Summarise the main points raised in the planning debate.

3 State how your class voted at the end of the exercise, and whether you agree with this 'democratic' decision.

Additional Exercises

1 Pair up these key terms with their correct definitions:

Terms	Definitions
A conurbation	are reasons why people migrate from one place to another.
A dormitory settlement	are South African settlements inhabited mainly by one ethnic group.
A honeypot	are usually created by the people themselves rather than large, well-established companies.
Informal sector jobs	
Literacy	
Malnutrition	describes the general well-being of a community.
Millionaire cities	have more than 1 000 000 inhabitants.
Pull and push factors	is a large, built-up area which includes a major city.
Quality of life	is a place in which people live but tend to work elsewhere.
Townships	is a tourist attraction which draws a large number of visitors.
	is the ability to read and write properly.
	is the lack of a balanced diet.

2 Pair up the following named places with their correct descriptions:

Names	Descriptions
Andersonstown	is a country park in Northern England.
Beacon Fell	is a mainly Catholic district in Belfast.
Runcorn	is an English new town.
Sha Tin	is a new town in Hong Kong.
Toxteth	is an inner city area in Liverpool.

3 State whether each of these statements is *true* or *false*:

a Most back-to-back housing dates from the nineteenth century.
b The early 1960s was the peak period of immigration into Britain.
c The present 'troubles' in Northern Ireland are largely the result of migration in the past.
d By-pass roads help through-traffic to avoid town centres.
e British supermarkets are much larger than American regional shopping centres.
f Rural-urban fringe areas have some of the features of both town and country.
g Perception is important because it helps people to decide how to make the best use of an area.
h China clay is mined in Cornwall.
i On topological maps, the distances along routes between settlements are always drawn to scale.
j The Forestry Commission was created in 1819.

4 Select the *most appropriate* ending to each of these statements:

a One example of a developed country is:

☐ Australia
☐ China
☐ Zambia.

b Developing countries tend to have high birth rates because:

☐ parents believe that many of their children will die before reaching adulthood
☐ of better medical care
☐ of improved transport facilities.

c A common *push* factor in rural areas is:

☐ biased reporting by the media (e.g. television)
☐ the availability of better housing in the cities
☐ the lack of non-agricultural employment.

d A primate city is:

☐ a capital city
☐ by far the largest city in a country
☐ where the Pope or a Prime Minister lives.

e Another name for the shatter belt in a city is:

☐ Central Business District
☐ commuter zone
☐ twilight zone.

f The type of urban zone from which outward migration is greatest is:

☐ inner residential area
☐ neighbourhood unit in a new town
☐ outer suburb.

g A type of land use which tends to cluster is:

☐ banks
☐ football grounds
☐ post offices.

h San Fransisco's Chinatown is sited:

☐ near the city centre
☐ on the edge of the city
☐ about 35 km from the city centre.

i Industrial estates are:

☐ areas of land set aside for new factories
☐ housing estates with some factories
☐ mainly in the Central Business District.

j Comprehensive redevelopment is:

☐ building new secondary schools.
☐ demolishing the old buildings in an area and replacing them with new ones.
☐ modernising old buildings to give them a new lease of life.

5 Complete the following statements as fully as possible:
a Gross National Product is a way of measuring . . .
b Developed countries are countries which . . .
c The population growth rate of a country is the difference between its . . .
d High-rise flats are now very unpopular because . . .
e A shanty town is a part of a large settlement which . . .
f Recent immigrants tend to live in our inner residential areas because . . .
g Rapid transit systems are built only in densely populated areas because . . .
h The number of corner shops in Britain has declined rapidly in recent years because . . .
i The main aims of the National Trust are to . . .
j Planning enquiries are often held in public to give people an opportunity to . . .

6 Suggest one kind of information *included in this book* which *should* be plotted by using each of the following types of graph:

a Bar graph.
b Divided bar graph *or* pie graph.
c Line graph.
d Population pyramid.
e Scatter graph.

Study the Ordnance Survey map extract on page 148 then answer Questions 7–10 which are based on it.

7 a What is the distance along the top of this map (in km)?
b What is the distance down the sides of it (in km)?
c What is the area of the map extract (in km^2)
d In which general direction are you travelling if you drive along the M20 motorway from Wrotham Heath *to* Leybourne?
e What is the height (in m) of the highest point on the map?

8 Write down the 4-figure grid reference numbers of the squares which have these village *names* in them:
a Addington **c** Leybourne **e** Ryarsh
b Birling **d** Offham

9 Identify the *main* land use at each of these 6–figure grid reference positions:
a 650593 **c** 675594 **e** 711592
b 658602 **d** 680552

10 Describe the *layout* of the built-up area which is south of the River Medway, and extends eastwards from Larkfield to the edge of the map. (Hints: consider land uses and different house-building periods; use key terms such as 'in-filling' whenever possible.)

11 Write 1–3 page essays to *discuss* the following themes. The last paragraph in each answer should say whether you agree what is stated – after considering all the evidence both for and against.
a 'Whether conflict in rural areas can be avoided in *heavily* populated countries.'
b 'Whether the quality of life of British *city-dwellers* has definitely improved since 1900.'
c 'Whether migrations (of all kinds) have created *more* problems than they have solved.'
d 'Whether *every* large British town should have its own green belt.'
e 'Whether our "new towns" have met *all* the needs of the people who have moved into them.'
f 'Whether developed countries should give *more* aid to those in the Third World'.
g 'Whether the general public should *always* be asked to comment on major planning proposals.'

Suggested Fieldwork Topics

Topics based on aims

To investigate population changes in . . . (name of region, district or settlement) *during the period* . . . (years from and to). Possible themes are population density, age/sex structure, ethnic composition. Use national cenuses and questionaire-based information on migration of individuals.

To identify the pattern of urban zones in . . . (name of town or city). Plot observations of land use, building styles and periods on street plan. Use additional information from local library's historical resources and interviewing residents and workers. House prices and types can be obtained from local papers and estate agents' property guides.

To identify the catchment area of . . . (name of country park). Interview visitors to discover where they live; plot this information on base map and in graphical form.

To identify areas of conflict within . . . (name of rural area). Obtain information on land uses in study area. Interview residents to identify areas of greatest conflict.

To assess the efficiency of the . . . (road or rail) *network in* . . . (name of region). Use the techniques described in Unit 7.4.

To discover the most common types of business in suburban shopping parades in . . . (name of town). Plot positions of parades on base map, then graph and discuss their range of businesses.

To investigate the distribution of recreational facilities in . . . (name of town). Use the techniques described in Study 10 on pages 65–67 of *Fieldwork Studies in Geography*.

Topics based on hypotheses

That . . . (name or grid reference of rural-urban fringe area) *is still more rural than urban in character*. Plot land uses on large-scale O.S. maps. Interview residents to obtain detailed information and *their* responses to the statement in the hypothesis.

That urban renewal in the . . . (district) *of* . . . (name of settlement) *has improved the quality of life of its inhabitants*. Base assessments on observations and interviews of residents.

That . . . (name of village or small town) *is a dormitory settlement*. Obtain information on employment available within study area and other settlements within commuting distance. Use a questionaire to discover work-places of people living in study area.

That the pattern of shopping facilities in . . . (name of town) *has changed considerably during the last* . . . (number of) *years*. Obtain details of businesses which have begun or ceased trading as well as those whose pattern of trading (e.g. range of goods for sale) has changed.

That the provision of services in . . . (name of rural area) *has declined during the last* . . . (number of) *years*. Consider public transport, schools, shops, garages, social facilities etc.

That the residents of . . . (name of town) *have an accurate perception of the* . . . (name of part of town's CBD or well-known rural area). Could be based on one of the questions in Unit 9.1.

That certain types of land use in . . . (name of town) *tend to cluster more than others*. Use the techniques described in Study 8 on pages 61–63 of *Fieldwork Studies in Geography*.

Glossary

accessibility How easily a place can be reached. Accessible places are easy to get to; inaccessible ones are not.

afforestation Planting trees on land previously used for other purposes.

annual population growth Rate at which a country's population increases over a whole year; the difference between birth and death rates, after allowing for international migration. A *decrease* in population is shown by a minus sign.

apartheid Official policy of the South African government which has led to the segregation of the different ethnic groups in that country.

back-to-back housing Small type of early Nineteenth Century terraced housing which shared a back wall with the dwellings behind it.

barrage Long dam across a river estuary or bay which creates a reservoir of fresh water.

Beta Index Number which shows how *directly* the places in a transport network are linked.

birth rate Average number of births in a year, per thousand people.

by-law housing Type of terraced dwelling which replaced back-to-back housing after 1875.

capitalist Describes a system of government which encourages private enterprise and the creation of personal weath.

census Population count.

Central Business District (CBD) Most central zone in a settlement, where land values are highest and many people work but few actually live.

clustering Tendency for similar land uses to group together within a town or city.

communist Describes a system of government which encourages the shared ownership of assests and a classless society.

commuter Person who travels some distance to work each day.

comprehensive redevelopment Form of urban renewal which involves demolishing old buildings and replacing them with new ones.

conflict Disagreement(s) arising when different activities take place within the same area.

conversationist Person who tries to maintain or increase the attractiveness of an area.

conurbation Very large built-up area which includes at least one major city.

convenience foods Foods which are specially packed or treated to preserve them and make them easier and quicker to prepare for eating.

counter-urbanisation (also **outward migration**) Migration of people away from central urban areas to the outskirts or rural areas beyond them.

country park Rural area set aside mainly for recreational use; much smaller than a national park.

death rate Average number of deaths in a year, per thousand people.

derelict Describes a building or an area which is no longer in active use and usually very unsightly.

developing country Country with a low per capita GNP and generally low standard of living. Most adult workers are farmers

developed country Country with a high per capita GNP and generally high standard of living. Usually heavily industrialised.

dormitory settlement Settlement in which many of the working people are commuters.

edge Stretch of route joining two nodes on a topological map.

enumeration district Large population census area having 2 000–8 000 people.

environment General term describing the nature of an area.

erosion Wearing away of the land surface.

ethnic group Group of people who share the same language, colour and culture.

expanded town New town built onto an existing settlement.

formal sector Regular employment in public services and well-established companies.

gentrification Form of urban renewal in which large old dwellings are bought by wealthy people and expensively modernised by them.

glaciated Shaped by ice.

green belt Mainly rural area around a large city in which new developments are strictly controlled; created to restrict urban sprawl.

Gross National Product (GNP) Total wealth created by a country in any one year; obtained by adding the value of all the goods it manufactures and services (e.g. shipping) it renders.

group of people Section of the community which has one or more features (e.g. age) in common.

high order goods Goods which are very expensive and usually bought only infrequently (e.g. furniture, jewellery). Low order goods are the complete opposite (e.g. bread, newspapers).

high-tech Describes an industry which is involved in the most up-to-date technology (e.g. the use of robots to do routine tasks).

honeypot Extremely popular tourist attraction.

hypermarket Very large modern shopping development which can provide an almost complete range of goods as well as many other types of facility.

industrial estate Land set aside for the building of groups of new factories.

in-filling Building on open land between two existing ribbon developments.

informal sector Jobs created by the workers themselves.

land intensive Describes any development which requires large amounts of land (e.g. hypermarkets, motorways).

linear Describes a settlement which is long and narrow.

literacy The ability to read and write. Literate people can do both; illiterate people cannot. Literacy rate is the percentage of people in a country who are literate.

localised clustering Clustering within a small area – often along a particular street.

malnutrition Un-balanced diet lacking certain types of goodness (e.g. calories and vitamins).

matrix Table with rows and columns; used to compare the accessibility of places in a transport network.

megalopolis Exceptionally large built-up area made up of a number of conurbations.

migration Moving home from one area to another. Emigration is moving *away* from the old area; immigration moving *into* the new one.

millionaire city City with at least 1 000 000 people.

multi-racial society Community having a number of different ethnic groups.

National Park Large, mainly rural area whose outstanding scenery and/or wildlife are conserved for public enjoyment.

neighbourhood unit Residential area within a new or expanded town having its own basic facilities and sense of identity.

new town Settlement which has been carefully planned as a whole unit and built within a short period of time.

node Place located on a topological map.

nucleated Describes the shape of a settlement which is generally rounded in outline.

outskirts The outer areas of a town or city.

overspill population 'Surplus' population moving out from old, overcrowded urban areas as a result of urban renewal.

parade of shops Small group of terraced shops within a residential zone offering only a limited range of low order goods.

per capita GNP The gross national product of a country divided by its population.

perception How our minds 'picture' an area.

pollution Harmful effect on the environment caused by human activity. Includes noise, vibration, dirt and toxic substances.

population density Number of people living in an area (usually written as . . . people per km^2). Many people live in densely populated areas; few people live in those which are sparsely populated.

population distribution The way in which the population of an area is spaced out within it.

poverty cycle Series of stages through which Third World countries pass because they lack the wealth needed to improve their quality of life.

primate city Largest city in a country which is considerably bigger than its closest 'rivals'.

pull factor Reason for moving *into* an area.

push factor Reason for moving *away from* an area.

quality of life General assessment of the well-being of a community and the area in which it lives.

rapid transit system Railway network within a conurbation; the inner sections are often routed underground.

recreation Activities undertaken for pleasure rather than profit.

renovation (also **rehabilitation**) Form of urban renewal in which old buildings are modernised.

ribbon development Building along a main road; produces a linear-shaped settlement.

rural Concerning the countryside.

rural-urban fringe Area around a settlement which shares the features of both town and country.

rural → urban migration Movement of people from the countryside into the towns.

rush hour Period in the day when traffic congestion is particularly severe.

segregation Deliberate separation of different activities, land uses or groups of people.

settlement Any built-up area, of any size.

shanty town Area of un-planned and often inadequate housing on the outskirts of a Third World city; populated mainly by immigrants from rural areas.

shatter belt (also **twilight zone**) Part of an inner residential zone greatly affected by urban renewal.

Shimbel number Number which shows the accessibility of a place in a transport network.

shopping hierarchy The way in which shopping facilities occur in order of size and range of goods offered for sale.

shopping mall Covered 'street' within a modern indoor shopping complex.

site Describes the position of the land on which a settlement has been built.

site and service scheme Plan by which a government provides financial help and guidance for new housing projects, but the people do most of the building work themselves.

situation The position of a settlement in relation to the large area around it.

smog A harmful mixture of *smoke* and *fog*.

Spearman Rank Correlation Coefficient Number showing how closely two sets of information are linked.

sprawl Outward growth of a settlement into the countryside around it.

stacking stalls Type of shopping facility found in the poorer districts of Third World cities in which two or more stalls are built on top of each other.

subsidies Ways of giving financial assistance. These include cash grants and tax concessions.

suburb Large area of modern housing built on the outskirts of a settlement.

suburban shopping centre Group of shops, often with other types of business as well, which is located some distance from the town centre and serves the residential area around it.

superstore Large, modern shop which sells a wide range of goods, including food. Most supermarkets are on one floor and have the warehouse type of construction.

tenement Type of high-density accommodation built in Scotland during the late nineteenth century.

tenure Conditions under which property is occupied (e.g. owner-occupied means that dwellings are owned by the people living in them).

Third World Collective term for all developing countries.

topological map Map not drawn to scale in which all transport routes are shown by straight lines.

township Large settlement in South Africa inhabited mainly by non-whites.

under-nourishment Not eating enough food to keep healthy.

urban Concerning built-up areas.

urbanisation Tendency for people to live in urban rather than rural areas.

urban renewal Measures taken to give an urban area a new lease of life.

urban zone Area within a town which has different land uses to the other areas around it.

ward Small population census area consisting of only one or two streets.

Global Case Study Locations

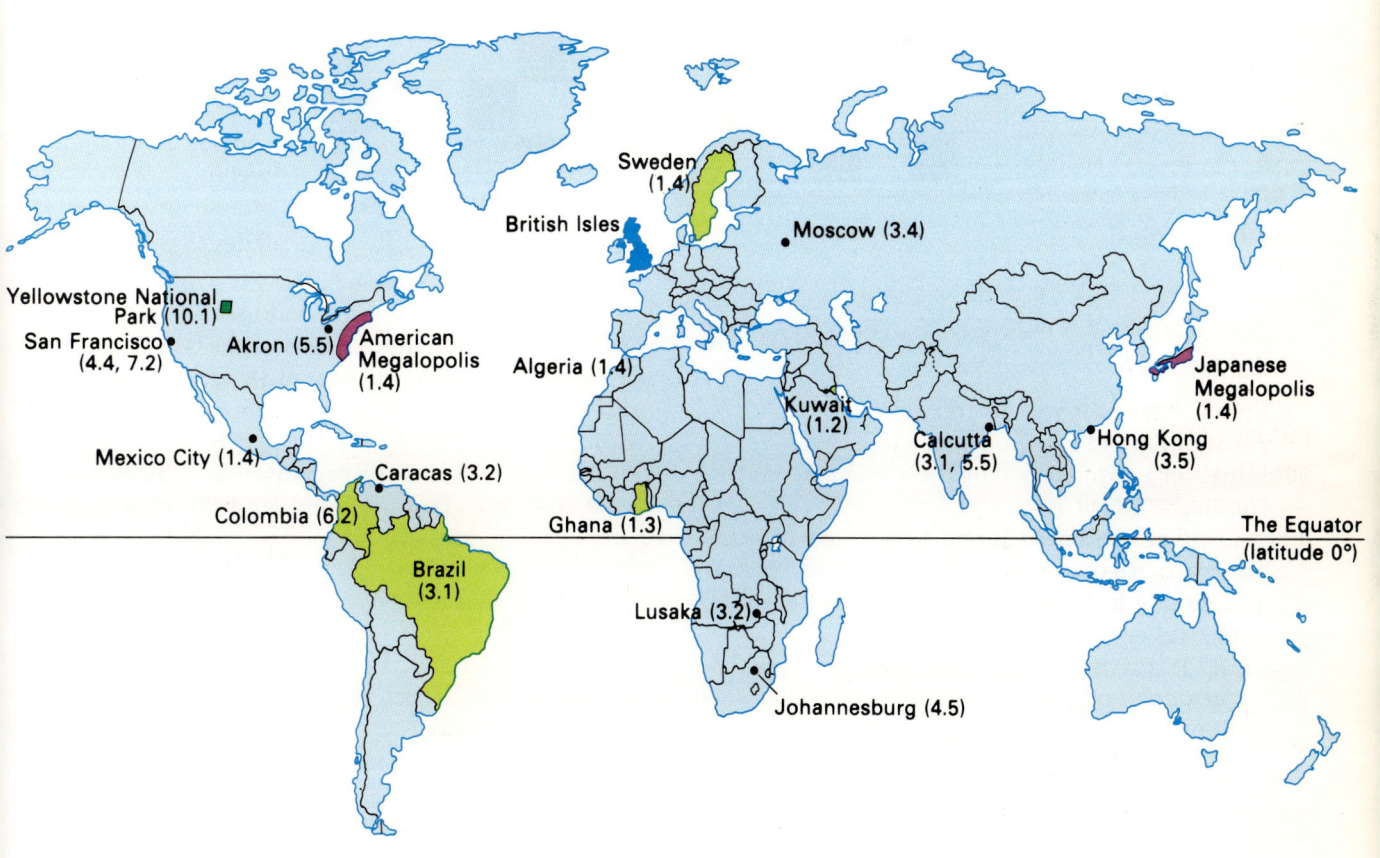

Sweden
(1.4)

British Isles

Moscow (3.4)

Yellowstone National
Park (10.1)

San Francisco
(4.4, 7.2)

Akron (5.5)

American
Megalopolis
(1.4)

Algeria (1.4)

Japanese
Megalopolis
(1.4)

Kuwait
(1.2)

Calcutta
(3.1, 5.5)

Hong Kong
(3.5)

Mexico City (1.4)

Caracas (3.2)

Colombia (6.2)

Ghana (1.3)

The Equator
(latitude 0°)

Brazil
(3.1)

Lusaka (3.2)

Johannesburg (4.5)

Crossword

Clues across

1 '... of life' describes the well-being of a community (7 letters). *Unit 1.1.*
4 British company owning many superstores (4). *Unit 5.2.*
8 Primate city of Northern Ireland (7). *Unit 4.3.*
9 Form of land use in CBD areas (5). *Unit 2.2.*
11 Initials of a term describing the annual income of a country (3). *Unit 1.2.*
13 ... settlements are long and narrow (6). *Unit 2.2.*
15 The two rush hour peak travelling periods are about breakfast and ... time each day (3). *Unit 7.1.*
16 Venezuela's chief source of income (3). *Unit 3.2.*
17 ... examples of a megalopolis have been described in this book (3). *Unit 1.4.*
18 '...-standard' describes inadequate housing (3). *Unit 2.3.*
20 General term for improving built-up areas (7). *Unit 2.5.*
23 Renovation schemes mean that only unsatisfactory and very ... buildings are demolished (3). *Unit 2.5.*
24 ... Transit Systems (5). *Unit 7.2.*
26 Spearman's ... Correlation Coefficient shows how closely two sets of information are linked (4). *Unit 6.1.*
27 Large wild animal in Yellowstone National Park (3). *Unit 10.1.*
28 '... er-occupiers' have bought the houses they live in (3). *Unit 2.3.*

Clues down

2 Refers to built-up areas (5 letters). *Unit 1.4.*
3 '... ing' takes place on vacant land between ribbon developments (6). *Unit 2.2.*
5 Housing areas near the edge of a town (7). *Unit 2.2.*
6 Special type of road in Runcorn New Town (8). *Unit 8.4.*
7 Initials of a voluntary organisation which owns and conserves buildings and land (2). *Unit 11.1.*
10 Immigrants are ... usually attracted to small settlements (3). *Unit 4.2.*
12 Describes many developing countries (4). *Unit 1.2.*
14 A route-system (7). *Unit 7.4.*
17 Developed countries are not of this world! (5). *Unit 1.2.*
19 Type of housing named after a type of Local Government instruction (5). *Unit 2.1.*
21 Where two or more edges meet (4). *Unit 7.4.*
22 Fashionable name for a road (4). *Unit 2.2.*
25 Type of business found in most suburban shopping centres; also the name of a river in northern Italy. *Unit 5.2.*

Index